Routledge Revivals

Competition for Wetlands in the Midwest

In the Midwest, wetlands can be seen as a nuisance to farmers as they can severely increase production costs. Wetlands are, however, a major part of ecology for migratory waterfowl and drainage of these wetlands could have dire consequences on the population of waterfowl as well as other wildlife. Originally published in 1971, this report attempts to break down the economic factors of competition for wetlands in Minnesota and surrounding areas in a policy-relevant way and to suggest new policy alternatives. This title will be of interest to students of Environmental Studies.

Competition for Wetlands in the Midwest

An Economic Analysis

Jon H. Goldstein

First published in 1971
by Resources for the Future, Inc.

This edition first published in 2016 by Routledge
2 Park Square, Milton Park, Abingdon, Oxon, OX14 4RN
and by Routledge
711 Third Avenue, New York, NY 10017

Routledge is an imprint of the Taylor & Francis Group, an informa business

© 1971 Resources for the Future, Inc.

All rights reserved. No part of this book may be reprinted or reproduced or utilised in any form or by any electronic, mechanical, or other means, now known or hereafter invented, including photocopying and recording, or in any information storage or retrieval system, without permission in writing from the publishers.

Publisher's Note
The publisher has gone to great lengths to ensure the quality of this reprint but points out that some imperfections in the original copies may be apparent.

Disclaimer
The publisher has made every effort to trace copyright holders and welcomes correspondence from those they have been unable to contact.

A Library of Congress record exists under LC control number: 74149240

ISBN 13: 978-1-138-95916-3 (hbk)
ISBN 13: 978-1-315-66075-2 (ebk)

COMPETITION FOR WETLANDS
IN THE MIDWEST

COMPETITION FOR WETLANDS IN THE MIDWEST:
An Economic Analysis

By Jon H. Goldstein

RESOURCES FOR THE FUTURE, INC.
Distributed by The Johns Hopkins Press
Baltimore and London

RESOURCES FOR THE FUTURE, INC.
1755 Massachusetts Avenue, N.W., Washington, D.C. 20036

Board of Directors:
Erwin D. Canham, *Chairman*, Robert O. Anderson, Harrison Brown, Edward J. Cleary, Joseph L. Fisher, Luther H. Foster, F. Kenneth Hare, Charles J. Hitch, Charles F. Luce, Frank Pace, Jr., William S. Paley, Emanuel R. Piore, Stanley H. Ruttenberg, Lauren K. Soth, P. F. Watzek, Gilbert F. White.
Honorary Directors: Horace M. Albright, Reuben G. Gustavson, Hugh L. Keenleyside, Edward S. Mason, Laurance S. Rockefeller, John W. Vanderwilt.

President: Joseph L. Fisher
Vice President: Michael F. Brewer
Secretary-Treasurer: John E. Herbert

Resources for the Future is a nonprofit corporation for research and education in the development, conservation, and use of natural resources and the improvement of the quality of the environment. It was established in 1952 with the cooperation of the Ford Foundation. Part of the work of Resources for the Future is carried out by its resident staff; part is supported by grants to universities and other nonprofit organizations. Unless otherwise stated, interpretations and conclusions in RFF publications are those of the authors; the organization takes responsibility for the selection of significant subjects for study, the competence of the researchers, and their freedom of inquiry.

This study is part of the RFF natural environments research program directed by John Krutilla. Jon H. Goldstein is a research economist in the Office of Research and Statistics, Social Security Administration. The charts were drawn by Federal Graphics.

RFF editors: Henry Jarrett, Vera W. Dodds, Nora E. Roots, Tadd Fisher.

Copyright © 1971 by Resources for the Future, Inc., Washington, D.C.
All rights reserved
Manufactured in the United States of America

Library of Congress Catalog Card Number 74-149240

ISBN 0-8018-1309-3

CONTENTS

	Page
Preface	ix
Acknowledgments	xiii
1. Introduction	1
A Formal Model	4
The Study	8
2. Drainage Cost and the Agriculture Sector	9
The Sample of Drainage Systems	10
The Technology of Drainage	10
Restrictions on Drainage Subsidies	12
Drainage of Temporary Wetlands	14
Estimation of Future Revenue Stream	16
Distribution of Temporary Wetlands under the Assumption of Zero Additional Cultivation Costs	19
Tile Drainage	21
Ditch Drainage	24
Distribution of Temporary Wetlands When Additional Cultivation Costs Are Nonzero	25
Analysis of Table 5	29
The Allocation of Permanent Wetlands	32
Statistical Results	35
Private Profitability of Reclaiming Permanent Wetlands under Conditions of Price Supports	37
Summary	38
Temporary Wetlands	39
Permanent Wetlands	39
Appendix to Chapter 2.—Tables; Per Acre Cost of Cultivation by Crop; An Apparent Paradox in Table 5	41

3. A Model for Evaluating Waterfowl and Their Environment	45
Invalid Evaluation Procedures	46
Fee Hunting: A Possible Market Value for Waterfowl	47
A Suggested Model for Evaluating Waterfowl and Wetlands	49
Appendix to Chapter 3.—Sampling Procedure and Description of the Data; Summary of Response Statistics; Questionnaire	57
4. Wetlands as Breeding Habitat	62
Methodological Design and Data Sources	64
The Contended Importance of Type 1 Wetlands	66
Analysis of the Waubay Ecological Data	68
Aggregate Analysis	75
Implications of Breeding Habitat Depletion upon Hunting Harvest	79
Economic Implications	84
Summary	85
Appendix to Chapter 4.—Environmental Differences at Waubay; Statistical Relationships between the Number of Prairie Ponds and the Location of the Waterfowl Population	87
5. The Distribution of Hunting Land in Minnesota	90
Allocation of Privately Owned, Permanent Wetlands	91
Rental Value of Minnestoa Hunting Land	93
The Influence of Drainage on Hunting Quality	96
Allocation of Publicly Owned Hunting Land	98
Summary	100
Policy Implications	102

TABLES

1. Per Acre Cost of Draining Temporary Wetlands 15
2. Percent of Land Devoted to Specified Crops 17
3. Present Value of the Income from an Acre of Drained,
   Temporary Wetland (Tile Region) 21
4. Present Value of the Income from an Acre of Drained,
   Temporary Wetland (Ditch Region) 25
5. Number of Years Before the Present Value of
   the Income Stream Is Identical with the Investment Cost 28
6. Present Value of an Acre of Reclaimed Land Evaluated
   at Free Market Prices with a Discount Rate of 8½ Percent 36
7. Present Value of an Acre of Reclaimed Land Evaluated
   at Support Prices with a Discount Rate of 8½ Percent 37
8. Minnesota Waterfowl Population Estimates
   and Band Recovery Data 82
9. Present Value of an Acre of Arable Land Minus the Cost
   of Reclaiming It 92
10. Per Acre Rental Value of Wetlands Necessary
    to Forestall Drainage 93

Appendix Tables

2A.1. Changes in Yields per Acre Resulting from Drainage 41
2A.2. Comparison of Estimated Free Market Prices
      and 1963 Support Prices for Selected Crops 41
2A.3. Range of Values for Annual Addition to Gross Revenues
      from an Acre of Drained Wetland 41
2A.4. Per Acre Cultivation Costs by Crop 42
2A.5. Uniform Cost of Cultivating an Acre of Output 43

PREFACE

The wetlands of the Canadian Prairie Provinces, extending into the United States in the pothole country of the Dakotas and Minnesota, provide a classic case of conflict in resource utilization. Wetlands represent potential nuisances at best and substantial increases in costs of agricultural production at worst for farmers on whose land they occur. The disadvantages of marshes and ponds for the individual farm operator encourage their drainage and conversion to cropland. At the same time, these wetlands provide a vital part of the ecology of migratory waterfowl, the principal wildlife resource associated with the wetlands. Thus the drainage of wetlands has an adverse, uncompensated impact on the waterfowl population along with adverse consequences for furbearers and other perhaps less significant wildlife species.

Such conflicts in resource use are not unusual. Indeed, the allocation of any resource to one use precludes its simultaneous use for incompatible alternative uses. This problem is the substance of economic choices in both production and consumption. In the case of typical allocative decisions, the prices in the market reflect, as a general matter, the marginal valuation that producers and consumers place on additional units of resources in alternative applications. This is the way in which the economy's allocative work gets done. In the case of the wetlands, however, a misallocation is likely because migratory waterfowl have no market prices, being fugitive resources subject to harvest under specified game laws and regulations. Accordingly, the value of waterfowl, and hence the value of wetland resources used in waterfowl production is not recorded in conventional market transactions on which the allocation decisions of farmers are based.

Given these conditions, there are grounds for believing that the incentives for wetland owners tend to be biased in favor of more extensive drainage than would be economic if all of the benefits and costs were properly reflected in the decisions to drain or retain the wetlands and the costs and gains were equitably distributed. Moreover, further incentives to drain are provided by the subsidies paid under the Department of Agriculture's Agricultural Conservation Program for aiding the drainage of temporary wetlands in the prairie pothole region of the United States.

With this general problem in mind, Jon Goldstein undertook to analyze the wetlands allocation issue while he was a graduate student at the University of Minnesota, and he has continued work on the problem for some additional time. By restricting his analysis to the wetlands of Minnesota, he was able to reduce the scale of the problem to manageable proportions, but his analysis is useful for understanding the problem not only in Minnesota but also in other parts of the prairie pothole region where similar conditions obtain. Goldstein was not entirely successful in his effort to fix the value of the migratory waterfowl production and harvest. He was successful, however, in taking the problem apart in many other ways and reassembling the pieces in a very informative and policy-relevant manner. As with much good research, the study provides its surprises. Conservationists' concern over the ACP assistance in draining temporary wetlands, while justified, may be directed toward the lesser part of the problem. Goldstein finds that perhaps a more serious problem arises in connection with the incentives which farmers have in some areas to drain *permanent wetlands* without ACP subsidies as a result of the agricultural price support program. As this latter program is not likely to be terminated for a variety of reasons, Goldstein suggests policy alternatives with a view to rigging the incentives to produce the correct allocative decisions without endangering other possible objectives of various agricultural programs.

That this study is timely is suggested by the general concern about the wetlands allocation problem reflected in proposed legislation which addresses, in part, the incentive structure for getting more efficient allocation of wetlands in the Central and Mississippi Flyways. As Jon Goldstein's study represents a careful, competent analysis by a disinterested party, the findings should be of great assistance in weighing the legislative and administrative considerations.

John V. Krutilla

ACKNOWLEDGMENTS

The gestation period for this work is exceeded in length only by the list of people and institutions to whom I am indebted for assistance. The research was initiated as a doctoral thesis at the University of Minnesota, but much of the writing was done while I was a professor at Pomona College and during one summer as a visiting research fellow at Resources for the Future. The project was originally funded by Resources for the Future, but additional financial support came from the University of Minnesota, Pomona College, the Haynes Foundation, Ducks Unlimited, the National Science Foundation, and the Western Data Processing Center at the University of California, Los Angeles. O. H. Brownlee, my adviser and teacher at the University of Minnesota, suggested the topic, criticized and improved innumerable formulations of the analytic model, and provided encouragement and guidance throughout the research and during my entire graduate career. For this I am eternally grateful. Scott Maynes, Professor of Economics at the University of Minnesota, helped design the statistical samples and techniques for analyzing the data, and made numerous substantive comments on the early drafts. I have had the benefit of many discussions with John Krutilla, without whose counsel the work would never have come to publication. Donald Bentley, Professor of Mathematics and Statistics at Pomona College, reviewed the statistical analysis and offered alternative approaches. Walter Crissey and Aelred Geis, Migratory Bird Populations Station, U.S. Bureau of Sport Fisheries and Wildlife, Laurel, Maryland; Robert Jessen, Minnesota Department of Conservation; and

Harvey Nelson, U.S. Department of the Interior, arranged for access to data on hunters and waterfowl, read and commented on the analysis, and protected an economist from his own naiveté in a world of biologists, ecologists, and ornithologists. Betty Duenckel at RFF and Earl Crabb at the University of Minnesota did yeoman service as computer programmers. The Soil Conservation Service opened their files to me and assisted in the sampling of drainage projects. Finally, Nora Roots struggled as editor to revive a manuscript that suffered severely from its origin as an academic tome. To all these and a legion of unnamed I express my appreciation. The responsibility for all remaining errors is mine alone.

Jon H. Goldstein

August 1970

COMPETITION FOR WETLANDS IN THE MIDWEST

1

INTRODUCTION

The allocation of wetlands between the agriculture and wildlife sectors has the classical characteristics of a resource distribution problem: one scarce resource with two alternative uses. The distribution of wetlands between the two sectors would be automatically and optimally determined by the operation of the market mechanism if the situation were purely competitive and free of characteristics which cause market failures. But subsidies in the agriculture sector and the presence of externalities and public goods properties prevent the market from functioning properly. The extent of drainage in the Upper Midwest during the postwar period together with crop surpluses has raised questions as to the social advisability of allowing the continued conversion of wetlands into arable land.

Wetlands come in a variety of depths, areas, and seasonal life spans. Some have surface water only after rain storms or the spring thaw and are usually dry during the rest of the season. Others form shallow lakes with depths up to ten feet. In the Upper Midwest these marshes, sloughs, potholes, and ponds constitute the major breeding and resting areas in the United States for waterfowl along the Mississippi Flyway. They also provide food and cover for other kinds of game birds and support muskrats and other aquatic furbearers. Without these marshlands the domestic waterfowl population would be reduced, and duck hunting in the Midwest would be affected by the disruption of migratory habits.

To the farmer, wetlands are often a costly nuisance. It takes longer and costs more to plant in irregular patterns around the wet spots. Machinery often becomes mired and may suffer costly damage. Finally, migrating waterfowl may attack the farmer's crops and add to his production costs.

The farmer may be able to obtain some return from the more permanent wetlands by marketing muskrat pelts or selling trapping rights. He may also lease waterfowl hunting rights, but competition from free public hunting areas often makes it difficult for him to obtain a price high enough to warrant maintenance of his property as wetlands. The temporary marshes, which hold water only during the spring, generate little or no revenue,[1] however, and consequently are likely targets for the trenching machines. Even as a hunter, the farmer is unlikely to be able to capture any significant portion of the output value of his wetlands. Assuming effective enforcement of state and federal regulations governing open seasons and bag quotas, the number of ducks that he may harvest is restricted. Since wetlands impose costs on the farmer, and present very limited opportunities for generating revenue, one can understand his wanting to drain such lands.

Drainage is made even more attractive by several government programs. The United States Department of Agriculture (USDA) grants technical and monetary assistance that defrays part of the capital cost of draining eligible land. Liberal interest rates are available to farmers who wish to borrow to finance their share of the capital expenditure. And finally, price support programs increase the expected revenue from the crops grown on the reclaimed land. Thus the USDA's programs both reduce the farmer's cost of drainage and increase the expected cash return from cultivation.

There are social costs involved in land reclamation, however, which are not considered by the farmer when he decides to engage in drainage. The wetlands are important as breeding

[1] Occasionally a farmer will be lucky, and extract a weak crop from his temporary wetlands.

and hunting areas, but the farmer finds it difficult to capture their value, and the hunter finds no adequate method of expressing his desires for preservation of the habitat. In addition to persons who hunt currently, there are undoubtedly some individuals who, although they may not hunt this year, would like to preserve the option to hunt in the future. No course of action is available through the market by which these people can exhibit such preferences. For most productive processes the nonavailability of a procedure by which option demand can be exercised is not a serious problem.[2] When people wish to consume some of the product, they simply bid for it in the market. If none is being produced at that time, production will resume if the price offered is attractive enough. But if it is difficult or impossible to resume production of a commodity once the process has been interrupted, disregarding option demand may lead to a misallocation of resources. In the case of waterfowl hunting, it may be technically impossible to provide the desired product if much of the habitat has already been converted to arable land. It is probably a bit extreme to make the analogy to the problem of regrowing the redwoods after the forests have all been logged, but the comparison illuminates the dimensions of the task. The difficult nature of reconverting arable land into waterfowl habitat cautions against allowing an unregulated market to determine whether waterfowl habitat should be retained or drained.

Clearly, a divergence exists between private and social values in the wildlife sector, with the benefits of waterfowl population maintenance accruing primarily to hunters, and the burden of the maintenance being borne largely by farmers. It is precisely because of this divergence that the market is likely to fail as a mechanism for allocating wetlands, and too few marshes are likely to be preserved.

[2] Option demand is like an insurance premium: it is the amount that people are willing to pay now in order to assure the availability of a commodity in the future. For a discussion of the concept see Burton Weisbrod, "Collective-Consumption Services of Individual Consumption Goods," *Quarterly Journal of Economics*, August 1964.

A Formal Model

There are two central questions at issue here: How severe is the misallocation of resources between the agriculture and wildlife sectors under current market conditions? And, if there were freely competitive prices in the farm sector and no subsidies to cover the cost of drainage, would more wetlands be converted into agricultural land than is socially desirable? In order to answer these questions it is necessary to evaluate the output of the land when it is devoted to the production of wildlife and recreational hunting, and to compare the net benefits from such production with the net benefits (under both current and competitive conditions) from committing the land to additional agricultural production.

Consider initially the value of the land to the farmer if he drains it, assuming competitive conditions in the agriculture sector and no drainage subsidies.[3] Let

dq_{it} = the post-drainage change in output per acre of commodity i in year t. (This is stated in differential terms because for very temporary wetlands a weak crop is often attainable from the land even though it is not adequately drained. For permanent wet areas it will be the case that $dq_{it} = q_{it}$.)

P_{it} = the price of commodity i in year t.

dC_{it} = the cost of producing dq_{it}, i.e., the gross value of the additional resources such as fertilizer, seed, storage and interest costs.

C_{dt} = the cost which is imposed upon the farmer in year t as a result of having one acre of wetland on his property. (That is, C_{dt} is the nuisance cost of wetlands, e.g., time and machinery expenses as well as waterfowl depredation of the crops.) From a social point of view this is part of the opportunity cost of allocating the land to the wildlife sector. From the farmer's point of view $dC_{it} - C_{dt}$ = the net cost of production of dq_{it}.

---

[3] The model is modified in chapter 2 to accommodate the market conditions currently encountered by farmers.

I_s = the full resource cost of draining an acre of wetland.
r = the interest rate of relevance.
n = the number of years it takes the drainage system to totally depreciate.

Making the assumption that all arguments are invariant with respect to time, we have

$$V_{ap} = \sum_{t=1}^{n} \sum_{i=1}^{m} \frac{p_i dq_i - dC_i + C_d}{(1+r)^t} \tag{1.1}$$

= the private, discounted value of the income stream from an acre of drained land. (C_d appears as an addition to the farmer's revenues in this expression, because this is an annual cost which he saves as a result of having eliminated one acre of wetland from his property. It should be noted that under the assumption that waterfowl have no value and there is no alternative use for the land, V_{ap} also represents the social value of an acre of reclaimed land.)

If $I_s \leq V_{ap}$, and if there were neither an alternative use for the land nor any external economy or diseconomy associated with its use as agricultural land, one could say that drainage is economically justified. But there is an alternative use for some wetlands, and hence an opportunity cost associated with reclamation. In order to determine the optimal allocation of wetlands at the margin, the discounted value of the net social benefits from reclamation must be compared with the discounted value of the net benefits derived from committing the marsh to wildlife production and recreational use. Let

R_t = the rental value of wetlands per acre per season of waterfowl hunting in year t.
W_t = the value of the ducks bred per acre of wetland in year t.
C_{wt} = the per acre cost of generating R_t. (That is, there may be some maintenance expenses associated with operating a hunting slough. The value of crops destroyed by waterfowl is not included in C_{wt}, since these are accounted for in C_{dt}.)

Again assuming that variables are constant over time, we have the expressions

$$V_{ws} = \sum_{t=1}^{n} \frac{R + W - C_w - C_d}{(1+r)^t} \qquad (1.2)$$

= the discounted value of the net social benefits from an acre of wetland, and

$$V_{as} = \sum_{t=1}^{n} \sum_{i=1}^{m} \frac{p_i dq_i - dC_i}{(1+r)^t} \qquad (1.3)$$

= the discounted value of the net social benefits from an acre of drained land.

The condition for determining the socially optimal allocation of land at the margin is:

if $V_{ws} > V_{as} - I_s$, leave the land in the form of wetlands; (1.4)
if $V_{ws} < V_{as} - I_s$, allocate the land to the agriculture sector.

(If an equality holds, it is a matter of indifference how the land is allocated.)

In other words, if for any particular parcel of land one estimated the social income stream from the wildlife sector (V_{ws}) and from the agriculture sector (V_{as}) and the capital cost of reclamation (I_s) and found that the discounted value of the net social benefits from the drained land less the cost of drainage ($V_{as} - I_s$) exceeded the discounted value of the net social benefits from the land as marsh land (V_{ws}), then the appropriate use of the land from society's point of view would be in the agriculture sector. If the inequality ran in the opposite direction, then the land should remain as habitat.[4]

---

[4] In defining the social value of land in the wildlife and agricultural sectors (V_{ws} and V_{as}) C_d has been included in the expression for the value of land in the wildlife sector, for C_d is one of the costs associated with retention of land as wetland. In effect, $\Sigma C_d/(1+r)^t$ has been moved from the right- to the left-hand side of the inequality in 1.4, thereby altering V_{ap} to V_{as}. This action merely results in the social value of the land in its alternative uses being properly expressed, and in no way changes the direction of the inequality in 1.4.

Unfortunately, this comparison of social values is not the criterion employed by wetlands owners in determining the distribution of their land. The relevant criterion for marsh owners is one involving private values, and private and social values do not coincide here. There is no divergence between private and social costs and benefits in the agriculture sector, but in the wildlife sector W is a social value which is not also a private value. The owner of marshland may be able to capture the rental value of wetlands for hunting purposes, but there is no way in which he can appropriate for himself the value of the migratory birds that are bred as a result of the existence of his land. Legally the marsh owner is restrained from harvesting all the birds that light on his property, and there is no mechanism available by which he can market the contribution that his land makes to the maintenance of the waterfowl population. This is because no one hunter would be willing to voluntarily pay a marsh owner in order to induce him to preserve his wetland as breeding land, for that hunter would have a probability near zero of shooting the ducks bred there. Thus, as far as the wetlands owner is concerned, the breeding component of the land value is zero ($W = 0$), and the private value of marshland (V_{wp}) is likely to be less than the social value (V_{ws}). Depending upon the habitat characteristics of the marsh, W could be large enough to make the market allocation of land inefficient.

Although the public goods characteristics of breeding land make it irrational for an individual hunter to pay wetlands owners not to drain their land, collectively such action is not unintelligent. Waterfowl hunters as a group might be willing to have a tax imposed upon themselves, the revenues being transferred to owners of breeding land in order to insure retention of such lands in the wildlife sector. An example of such collective activity is Ducks Unlimited, a well-known organization financed by hunters' contributions, which purchases and maintains breeding land in Canada in order to guarantee larger waterfowl populations. Voluntary group action is a good deal more successful than the atomized attempts of individuals, but the dimensions of the resulting program are still likely to fall far short of reflecting the desires of the entire hunting community, because hunters can participate in the fruits of Ducks Unlimited's activities without donating to its treasury. It would thus be hazardous to rely upon this or similar organizations to preserve the socially appropriate amount of waterfowl habitat.

The Study

The body of this study is devoted to the evaluation of the variables in equations 1.2 and 1.3 and the determination of the direction of the inequality in 1.4. The structure of equations 1.2, 1.3, and 1.4 is modified where necessary to account for the existence of noncompetitive conditions or additional divergencies between private and social values. The principal area of investigation is Minnesota, but the analysis has implications for the entire Midwest and for the Prairie Provinces of Canada. Because of differences in wetlands types and the soil compositions in different regions, no one universal solution emerges from the analysis. Results are influenced by the area in which the wetland is located, its durability, and its proximity to other marshes. The discussion in chapter 2 is confined to the agriculture sector, with estimates being made of the capital expenditure required for drainage and of the value of the output forthcoming under both competitive and price subsidy conditions. Chapter 3 considers alternative procedures for evaluating waterfowl according to the preferences of duck hunters. Wetlands as breeding habitat is the topic of chapter 4, with emphasis being placed on the effect which reclamation has upon the size of waterfowl populations and the concomitant effect upon hunting quality. Chapter 5 deals with the distribution of hunting land, and includes an estimate of the rental value of wetlands for hunting purposes and a discussion of deviations from optimality resulting from the existence of entry-free public hunting preserves. Finally, policy suggestions are made for those situations where the private allocation of land differs significantly from the social optimum.

2

DRAINAGE COST AND THE AGRICULTURE SECTOR

Two central questions were posed in the introductory chapter: How severe is the misallocation of resources between the agriculture and wildlife sectors under current market conditions? In the absence of any subsidies to the agriculture sector, would an excessive amount of reclamation still occur? In order to resolve these questions and evaluate 1.4, the condition for determining the optimal allocation of the land, the importance in the wildlife sector of various types of wetlands must be investigated, and some determination of the opportunity cost involved in draining them must be made. In this chapter, however, the attack upon these problems is initiated by introducing the simplifying assumption that wetlands in their natural state are useful only for hunting purposes.[1] This assumption permits us to temporarily set W, the value of the breeding component of wetlands, at zero in 1.2; to ignore the major divergence which exists between private and social values in the wildlife sector; and to concentrate exclusively on some important problems in the agriculture sector. Would the drainage of wetlands still be privately profitable if all subsidies to agriculture (price supports for output, capital grants for drainage facilities, and low interest rates

---

[1] As we shall see, this assumption is not unrealistic for several wetland formations. Isolated temporary potholes not near more permanent habitat or large complexes of type 1 marshes like those found in southern Minnesota have no waterfowl value. Even isolated durable potholes may have very limited value for breeding purposes. See footnote 13, p. 67.

on loans) were eliminated? To what extent does each of these individual subsidies influence and encourage drainage?

Wetlands vary in depth, durability, and ecological characteristics. The distinctions are important because the economic feasibility of reclaiming land is dependent upon these characteristics and the location of the wetland, and because wetlands of different types have differential significance in the wildlife sector. Consequently, the differences between wetland types in each geographic location are carefully maintained throughout the analysis.

The Sample of Drainage Systems

In order to estimate the private discounted value of the income stream in agriculture (V_{ap}) and the full resource cost of drainage (I_s), a sample of drainage systems installed in 1963 in two Minnesota counties was drawn from the local offices of the Soil Conservation Service (SCS). Because of the time-consuming procedure involved, the sample was necessarily small (thirteen projects from Blue Earth County and fourteen from Stevens County); it was therefore felt that a more efficient estimate could be obtained by use of a stratified proportional sample rather than an unrestricted, random sample. The criterion for stratification was the size of the drainage system in acres drained. That is, the proportion of drainage systems of size x acres included in the sample is intended to be identical with the relative number of drainage systems of size x in the universe.

Two methods of drainage are prevalent in Minnesota: ditch and tile systems. Ditch drainage of agricultural land is confined primarily to the northwestern and west-central portions of the state (the Red River Valley region), while tile drainage is concentrated in the less porous soil areas of south and southwestern Minnesota. Stevens and Blue Earth Counties are representative of the drainage systems located in these two regions respectively.

The Technology of Drainage

A tile drainage system is an underground network of loosely connected cement pipes. Excess or gravitational soil water seeps through the connections and flows through the pipeline to an outlet. Such a system is installed when the soil composition is such that subsurface drainage is mandatory.

> In a poorly drained soil, the gravitational water stands near the surface during the spring season when the root systems are being formed. Crop roots cannot grow and live in the absence of air. Hence they do not penetrate down into this portion of the soil in which the gravitational water stands for any length of time. The result is that in such soils the plant roots are confined to the shallow upper layer. When dry weather continues for any length of time, this top layer of soil dries out, and the growth of the crop is seriously retarded, if not stopped entirely, for want of sufficient water.[2]

Similarly, crops are frequently drowned on such soils when water from a shower fails to drain off, and inundates the area.

Open ditch drainage is employed primarily to eliminate standing water or to accommodate rapid surface runoff. It is technically possible to utilize the relatively cheaper open ditch method to remove subsurface water, but the necessary number of ditches makes this technique impractical. Land which lies 100 to 150 feet on either side of an open ditch can be effectively drained, the water draining into the ditch due to gravitational force. (The lateral distance of effective drainage is dependent upon the depth of the ditch.) However, use of this procedure for draining a field with a subsurface water problem would mean sacrificing a considerable area of potentially arable land, the ditch itself being nontillable. In addition, the area would be scarred so thoroughly with ditches that cultivation costs would be greatly increased.

Standing water can be the result of two alternative soil configurations. An impermeable subsoil may form a basin or pothole through which water drains very slowly, or a high water table may have saturated the soil to such an extent that surface water is evident. In the case of a high water table an outlet ditch must be installed initially to drain off the surface water

---

[2] James A. King, W. S. Lynes, and Vernon H. Fobes, *Tile Drainage*, Mason City Brick and Tile Co., 1946, p. 31.

and lower the water table sufficiently to permit tile placement. A ditch may be sufficient for permanent drainage of a pothole if the impermeable soil layer lies close to the surface. Otherwise, a tile system will be needed in conjunction with the ditch outlet.

Restrictions on Drainage Subsidies

Since the passage of Public Law 87-732 in October 1962, drainage projects involving permanent wetlands in North Dakota, South Dakota, and Minnesota have not been eligible for subsidy or technical assistance from the Agricultural Stabilization and Conservation Service (ASCS) or the SCS. The law does not prohibit the landowner from converting his permanent wetlands into arable land. It merely makes the drainage of such lands ineligible for technical and monetary assistance from the ASCS or the SCS, and in so doing it makes their drainage much less attractive.

All requests for drainage assistance on type 1 or "temporary" wetlands[3] must be referred to the Fish and Wildlife Service (FWS) for consideration and authorization. The intent of this law may have been to restrict subsidized drainage to type 1 wetlands which in the judgment of the FWS have insignificant value as wildlife habitat, but its provisions are too weak to accomplish this. Under Section 16A of this law an FWS finding that drainage will be detrimental to wildlife preservation must

---

[3] The following official definitions appear in *Wetlands of the United States*, U. S. Department of the Interior, Fish and Wildlife Service, Circular 39, 1956, pp. 20–25.

Type 1.—Seasonally flooded basins.... The soil is covered with water, or is waterlogged, during variable seasonal periods but usually is well drained during much of the growing season....

Type 3.—Inland shallow fresh marshes.... The soil is usually waterlogged during the growing season; often it is covered with as much as 6 inches or more of water. It should be noted that these marshes may contain water until midsummer, at which time they may dry up completely or remain waterlogged for the rest of the season. It is unlikely, however, that surface water will still be evident after midsummer. In dry years these areas are sometimes arable.

Successively higher classificatory numbers indicate deeper, more permanent marshes. Classification number 2 is no longer used.

be issued within 90 days after the application for drainage assistance has been filed; otherwise, assistance may be granted. The FWS claims that it has inadequate staff to process all the applications within the 90-day limitation. Thus, some projects may escape scrutiny. A ruling by the FWS that the land does have wildlife importance is not sufficient to prevent drainage assistance. The FWS must also extend an offer to purchase the land within one year. Budget restrictions may prevent acquisition within one year. Finally, if within five years the property owner does not find an FWS offer acceptable, the USDA may subsidize the drainage project. Thus, even for land that is valuable as wildlife habitat, the law may only temporarily prevent public assistance for reclamation.

The ASCS also maintains a self-imposed restriction on eligibility that is designed to limit subsidies to land which is already undergoing cultivation:

> This practice [subsidization of drainage] is not applicable to land other than that devoted to the production of cultivated crops during at least 3 of the 5 years preceding that in which the practice is applied, or to pasture or hayland seeded during the previous 5 years and which was devoted to the production of cultivated crops for at least 1 of the past 5 years.[4]

Thus, it is not at all clear that P.L. 87-732 has had any influence upon the type of land which receives drainage subsidies. If the ASCS has been meticulous in the enforcement of its own stated restrictions on drainage assistance during the last decade, it can be argued that the law is largely superfluous.[5]

[4] *Agricultural Conservation Program: Handbook for 1964, Minnesota*, U.S. Department of Agriculture, Agricultural Stabilization and Conservation Service, December 1963, pp. 28–30.

[5] Since 1955 the ASCS has maintained that "the purpose of the program is to help achieve additional conservation on land now in agricultural production rather than to bring more land into agricultural production. Such of the available funds that cannot be wisely utilized for this purpose will be returned to the public treasury." *1955 Agricultural Conservation Program, National Bulletin*, U.S. Department of Agriculture, Agricultural Stabilization and Conservation Service, July 1954, sec. 11101.601.

Drainage of Temporary Wetlands

If, as seems likely, there are no external economies associated with agricultural production or consumption, any limitation on subsidies that encourage drainage improves the current allocation of resources. To this extent P.L. 87-732 and the self-imposed restrictions of ASCS represent desirable regulations.

However, they do not guarantee an optimal allocation of land between the wildlife and agricultural sectors, and this is the issue of central concern to us. The question still remains as to how the land would be distributed under competitive conditions and in the absence of any artificial restrictions. Would the market-determined distribution be optimal? For this, the distribution implied by our sample estimates under simulated competitive conditions must be examined.

The sample of drainage projects was chosen entirely from the files of the SCS, and therefore includes no drainage systems of permanent water areas. To the extent that drainage of permanent wetlands requires more resources than drainage of a temporary marsh of comparable acreage, our estimate of per acre investment cost undervalues the cost of such drainage systems. This problem is dealt with later in the chapter. For the moment we shall confine our attention to temporary wetlands.

For each of the systems included in the sample the following variables are known:

I_p = the private cost of draining an acre of marshland, i.e., the cost of installation to the wetlands owner,

S = the per acre monetary subsidy made by the ASCS toward the cost of installing the system,

E = the engineering cost associated with the drainage of an acre of wetland,[6] and

I_s = the full resource cost of draining an acre of wetland
    = $I_p + E + S$.

The results of the sample appear in table 1.

---

[6] Neither the SCS nor the ASCS price the technical assistance which they provide to the owner of the drainage system. Most of this assistance comes in the form of surveying, a service which the wetland owner would have to purchase from a private contractor in the absence of the government's aid. It was estimated that these services add approximately 10 percent to the total cost of a drainage project.

TABLE 1. PER ACRE COST OF DRAINING TEMPORARY WETLANDS

Cost	Tile drainage Mean	Tile drainage Standard deviation	Tile drainage Range	Ditch drainage Mean	Ditch drainage Standard deviation	Ditch drainage Range
Full cost (I_a)	$157.49	$33.39	$123.54–228.41	$49.68	$27.34	$16.05–111.42
Engineering cost (E)	14.31		11.23–20.76	4.51		
Subsidy (S)	50.15		39.50–62.59	15.49		
Private cost (I_p)	93.02	24.24	68.34–147.42	29.67	16.78	9.72–67.52
Private cost as percent of full cost (I_p/I_a)	59%		54–64%	59%		49–61%

Estimation of Future Revenue Stream

In order to evaluate the revenue stream forthcoming from the reclaimed land ($\Sigma p_i dq_i$ from 1.1), estimates were required for each commodity of the change in output per acre which occurred as a result of drainage, the distribution of the acreage among the various commodities, and the free market prices used to evaluate the additional output.

In estimating a representative crop distribution, it was assumed that reclaimed land would be allocated among the various commodities in exactly the same proportions as the land currently under full cultivation. Let

α_{ij} = the percentage of land devoted to commodity i on farm j, and

x_j = the number of acres of land reclaimed by the drainage system on farm j.

Estimates of these variables were provided by SCS field workers familiar with each of the drainage systems in the sample. It follows that

$a_{ij} = \alpha_{ij} x_j$ = the number of acres of reclaimed land on farm j devoted to crop i.

Forming a weighted average, we define

$$\beta_{i1} = \frac{\sum_j a_{ij}}{\sum_i \sum_j a_{ij}} = \frac{\sum_j \alpha_{ij} x_j}{\sum_j x_j}$$

= the expected percentage of reclaimed land devoted to crop i.

The values for β_{i1} appear in table 2. These values have been calculated entirely from data collected in conjunction with the sample of drainage systems. Since the criterion for selection of the elements in the sample was their relative size in acres, there is no a priori reason for considering that the crop distribution associated with the farms included in the sample is an unbiased estimate of crop distributions in general. Consequently, two additional estimates of crop distributions were computed. One

TABLE 2. PERCENT OF LAND DEVOTED TO SPECIFIED CROPS

	Tile drainage region			Ditch drainage region		
Crop	β_{i1}	β_{i2}	β_{i3}	β_{i1}	β_{i2}	β_{i3}
Corn	46%	45%	47%	48%	38%	11%
Soybeans	32	33	24	10	13	6
Wheat	9	4	2	3	4	13
Oats	10	10	14	16	23	27
Hay	3	7	12	14	10	25
Barley				8	12	18

Source: St. Paul files of the ASCS regarding the extent of drainage in each county, and *Minnesota Agricultural Statistics, 1963*, USDA and Minnesota Department of Agriculture, March 1963.

(β_{i2}) was the relative number of cultivated acres devoted to commodity i in the counties where the samples occurred (Blue Earth and Stevens), and the other (β_{i3}) was the same calculation for the set of counties where drainage was most prevalent (counties where more than 400 acres were drained in 1963).

One problem arises as a result of employing this procedure for estimating β_i. All three of the alternative crop distributions computed were derived from current rotation practices, and hence are dependent upon current output and factor prices. For a different price vector the present distribution of cropped land may not correspond to the profit-maximizing distribution. Since the output of the reclaimed land in the sample will be evaluated at estimated free market prices, not at current support prices, the analysis is vulnerable at this point. No means were available to accommodate this criticism.

It was beyond the scope of this study to undertake a controlled experiment to determine output changes resulting from drainage. No such study analyzing yield changes from drainage has been done for Minnesota, and the few available studies done for other regions proved inapplicable. Consequently, complete reliance was placed upon information supplied by SCS field workers acquainted with the productive characteristics of the area and with the particular farms in the sample. For each system in the sample pre- and post-drainage output estimates were provided.

An average change in yield per acre for each crop cultivated for both types of drainage systems was estimated as follows:

$$\overline{dq_i} = \frac{\sum_j a_{ij}(dq_{ij}/dD_j)}{\sum_j a_{ij}},$$

where a_{ij} is as previously defined, and

$dq_{ij}/dD_j =$ the additional per acre output of commodity i on farm j as a result of installing a drainage system on farm j.

The results appear in table 2A.1 in the appendix to this chapter (see p. 41).

The tabled values for dq_i were calculated for temporary wetlands. A statistic is also needed to measure the change in yield when permanent wetlands are reclaimed. Let

$q_{ij} =$ the post-drainage output per acre of crop i on system j.

For permanent wetlands dq_{ij}/dD_j is defined to be q_{ij}. Hence, for permanent wetlands, $\overline{dq_i}$ becomes $\overline{q_i}$, where $\overline{q_i}$ is defined as

$$\overline{q_i} = \sum_j a_{ij}q_{ij} / \sum_j a_{ij}.[7]$$

Two different price vectors were used to evaluate the agriculture outputs. One is a set of prices estimated under the assumption of free market conditions, i.e., no direct price supports or surplus disposal and storage programs. The other set of prices is

---

[7] The same objection can be raised regarding $\overline{dq_i}$ and $\overline{q_i}$ as was voiced against β_{i1}. That is, the values for $\overline{dq_i}$ and $\overline{q_i}$ have been determined from data collected as an adjunct to the sample of drainage systems and their costs, and the elements in this sample were not chosen because of their representativeness in terms of crop yields. We are on less precarious ground here than we were with β_{i1}, however. It is fairly certain that use of one of the published estimates of crop outputs for Minnesota would result in undervaluing $\overline{dq_i}$ and $\overline{q_i}$. Available published data, which provide estimates of per acre yields by crop by county, are averages and include all the cultivated land within a county. But drained wetlands are very likely to fall in the upper end of the output distribution. Consequently, the measures employed here are preferable, despite the small sample sizes involved in their derivation.

the 1963 support prices for the counties from which the sample was drawn. Support prices in these counties differ only slightly from those in the other principal drainage counties. The price vector data appear in table 2A.2, and the derived values for the annual addition to gross revenues from an acre of drained land are found in table 2A.3 (see appendix to this chapter).

Distribution of Temporary Wetlands under the Assumption of Zero Additional Cultivation Costs

Having estimated the gross revenue stream forthcoming from an acre of reclaimed land, we are now in a position to determine the optimal use of temporary wetlands which are interspersed with well-drained, fully arable land. The topography here has a swiss cheese appearance, the holes being ephemeral potholes. It is the usual practice in such a topography to make an attempt to cultivate the poorly drained areas along with the balance of the land. Consequently, there is little change in cultivation costs after the land has been adequately drained. Although fertilizer, seed, storage, and interest costs will increase, they will be offset by reductions in machinery, time, and labor expenses.

Because temporary wetlands are dry in the fall, they have no value for hunting purposes (i.e., $R = 0$ in expression 1.2). Since it was assumed earlier that wetlands have no value for breeding purposes (i.e., $W = 0$ in 1.2), the observation that $R = 0$ means that for purposes of the present analysis temporary wetlands have no alternative use in the wildlife sector, and consequently, as pointed out in the definition of V_{ap} (1.1), there is no divergence between the private and social values of the land. Under the assumptions imposed, if free market prices result in drainage being privately profitable, then the *socially optimal* use of the land is in the agriculture sector. Consequently, expression 1.4, the condition for determining the socially optimal allocation of the land, degenerates into

$$\text{if } V_{ap} < I_s, \text{ leave the land in the form of wetlands;} \\ \text{if } V_{ap} > I_s, \text{ allocate the land to the agriculture sector.} \quad (2.1)$$

The work in the preceding section results in the expression for V_{ap}, the discounted value of the income stream from draining an acre of land, being modified to the slightly more complex expression

$$V_{ap} = \sum_{t=1}^{n} \sum_{i=1}^{m} \frac{\beta_{is}(p_i dq_i - dC_i) + C_d}{(1+r)^t},$$

where a range of values for V_{ap} is generated by choosing different values for s, and where dC_i is the cost of producing $\overline{dq_i}$, i.e., the cost of the *additional* output, and C_d is the cost saving annually as a result of draining one acre.

The observation that the additional cultivation costs incurred after reclamation will be quite small[8] may be symbolized by setting $\Sigma_i \beta_{is} dC_i - C_d = 0$. Thus the expression for V_{ap} reduces to

$$V_{ap} = \sum_{i=1}^{m} (p_i \beta_{is} \overline{dq_i}) \left(\frac{1 - (1+r)^{-n}}{r} \right),$$

where $[1 - (1+r)^{-n}]/r$ is the discounted value of $1 per year for n years. The range of values for $\Sigma_{i=1}^{m}(p_i \beta_{is} \overline{dq_i})$ appears in table 2A.3 (see chapter appendix).

A relevant discount rate and time period over which the stipulated income stream is forthcoming must now be specified. For the ditch systems a periodic maintenance cost[9] was subtracted from gross revenues in the calculation of present value, the assumption being that this ensures the effectiveness of the ditch in perpetuity, i.e., $n = \infty$. The useful life of a tile system varies widely; a range of 20 to 50 years was assumed.[10]

Two alternative discount rates were adopted. Five percent loans are available to the wetlands' owner (with an SCS ap-

---

[8] That is, although some costs will increase ($\Sigma_i \beta_{is} dC_i$), they will be offset by the elimination of the nuisance cost associated with the presence of wetlands (C_d).
[9] One-third of the original investment every seven years. Source: SCS, Morris, Minnesota.
[10] Philip Manson, Department of Agricultural Engineering, University of Minnesota, St. Paul; and SCS, St. Paul, Minnesota.

proved project) from Farmers' Home Administration (FHA), but under free market conditions the owner would face higher commercial rates. Eight and one-half percent was assumed as representative of commercial rates.[11]

Tile Drainage

Consider (from table 2A.3) the lowest estimate of an annual addition to income from an acre of drained (temporary) wetland. This is $\Sigma_i p_i \beta_{is} \overline{dq_i}$ evaluated at free market prices, and it equals $33.01. The entries in table 3 are the discounted value of this income stream at $r = 8\frac{1}{2}$ percent after n years.

TABLE 3. PRESENT VALUE OF THE INCOME FROM AN ACRE OF DRAINED, TEMPORARY WETLAND (TILE REGION)

No. of years (n)	Rate of interest (r) $8\frac{1}{2}\%$
6	$150.31
7	168.96
.	.
.	.
.	.
10	216.59
11	230.05

It is evident from a comparison of table 3 with the estimates of the installation cost of a drainage system in table 1 that under the assumption of zero additional cost associated with cropping drained land, tile drainage proves quite lucrative, i.e., $V_{ap} > I_p$. What is more, the conclusion here that tile drainage is privately profitable is quite conservative, because V_{ap}, as calculated above, has an inherent downward bias. It was calculated using (1) the

---

[11] The choice of 8½ percent as the relevant commercial rate was influenced by the results of a recent study which took as a measure of the return on capital in the agriculture sector the ratio of gross rent paid to the value of rented farms. It was found that this ratio ranged from an average of 5½ percent in Ohio to 8½ percent in Wisconsin. This, together with the fact that 8½ percent is closer to the commercial rate, accounts for the choice of 8½ percent as an upper bound for r. Source: Arnold C. Harberger, "The Interest Rate in Cost-Benefit Analysis," *The American Economy*, ed. Jesse W. Markham (New York: George Braziller, 1963), p. 233.

lowest estimate of a per acre, annual income stream under free market prices, and (2) an 8½ percent interest rate, a rate that exceeds that available to farmers. Even with this downward bias, V_{ap} exceeds the mean I_s after seven years, and it exceeds the largest I_s value in the range of our sample after only eleven years. A conservative estimate for life expectancy of a tile drainage system is twenty years. Hence, even when the owner of wetlands must bear the burden of unsubsidized conditions in all markets (competitive prices for agricultural produce, commercial interest rates, and the full investment cost of drainage), it is still profitable to engage in tile drainage.

This result deserves additional explanation. It was assumed that the temporary wetlands have no alternative use in the wildlife sector. If it is assumed further that there is no greater social return to be attained by allocating capital to any other sector, then $V_{ap} > I_s$ implies that the reclamation of these wetlands under free market conditions is a socially desirable use of resources. In the face of current commodity surpluses and a reserve of Minnesota land in the Soil Bank, this inequality does not imply that it is efficient to devote current resources to the drainage of marshland and the production of additional output. Nor does it imply the opposite, i.e., that such drainage is currently nonoptimal. To determine whether it is desirable to devote current resources to drainage, it is necessary to evaluate V_{ap} at the price that would clear the market for the current level of output, not at the price that would clear the free market level of output.

Some elaboration may help to clarify this point. Consider figure 1, where p_s is the current support price for commodity i, p_f is the free market equilibrium price, and p_c is the market-clearing price for the current output, q_s. If V_{ap} evaluated at p_c were to exceed I_s, then the use of current resources to reclaim land would be optimal despite the existence of a large quantity of fallow land in the Soil Bank. This would be so, because the discounted value of the output evaluated at current market-clearing prices (i.e., the value of the land once drained) would exceed the investment cost necessary to reclaim the land.

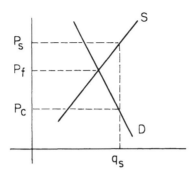

Figure 1.

This conclusion may elicit some criticism. Some may feel that farmers could increase their endowment of land more efficiently by transferring land in the Soil Bank back to the agriculture sector instead of converting wetlands to arable land. Admittedly, the former has a zero investment cost and the latter a positive one. However, drained land usually has above-average productivity, while land in the Soil Bank is likely to have been committed to this fate largely because of its marginal nature. In addition, our result rests on the assumption that no additional variable costs will be incurred when the drained land is cultivated. This would not be the case with the reserve of land in the Soil Bank. Hence, even in the presence of the current excess supply in the agriculture sector it may still be efficient to undertake some drainage projects. We were unable to estimate p_c, however, and thus no judgment could be made on this point.

Returning now to the situation where V_{ap} is evaluated with free market equilibrium prices, some interesting results follow from the conclusion that $V_{ap} > I_s$. The accusation that the subsidies granted by the USDA cause more land to be reclaimed than would have been the case under competitive conditions has been severely weakened. These subsidies did not make the difference between profitability and nonprofitability of draining any of the temporary wetlands in our sample. Given normal returns on capital elsewhere, the decision to install any of the tile systems in our sample should not have been influenced by the federal subsidies.[12] A tile drainage project with a twenty-year life would have to involve an investment of $312.38 per acre to be marginal under competitive conditions. This figure exceeds by 35 percent the most expensive investment project included in our sample. It appears, then, that under the assumption of zero additional cost for growing output, the federal programs influence few, if any, tile drainage projects.[13]

It is interesting to note just how lucrative tile drainage of temporary wetlands is under the current subsidization scheme. Using the lowest estimate for the annual addition to net revenue evaluated at support prices, and letting $r = 5$ percent (the interest rate available through the FHA), it can be shown that the investor recovers his capital (I_p) within three years if the system is of average expense, and within four years if the investment outlay is as costly as the most expensive system in the sample.

Ditch Drainage

Similar results are obtained if the same analysis is performed for ditch drainage. Choosing (from table 2A.3) our minimum estimate of an annual addition to income evaluated at free market prices ($21.39), we find that with a discount rate of 8½ percent, V_{ap} exceeds the mean, per-acre, drainage cost after three years, and the maximum I_s in the sample after eight years.

---

[12] This is not to say that these subsidies did not induce more wetlands to be drained than would have been drained in their absence. Returns on all other possible investments were undoubtedly not normal, and consequently the subsidies did alter the profitability of investment in drainage facilities relative to all other ventures. Since capital restraints are such that not all profitable projects can be undertaken, an increase in the relative profitability of drainage projects is very likely to result in more resources being devoted to reclamation.

[13] The prices that we postulated for the agriculture sector in obtaining the above results were supposed to be free market, equilibrium prices. But we have shown that more land would move into the agriculture sector via investment in drainage projects if these prices confronted producers. There is no inconsistency here with the postulated equilibrium prices if we assume that the additional acreage that would be reclaimed is small relative to the total amount of acreage under cultivation.

The lucrative nature of drainage under current, subsidized conditions is evident from a comparison of table 4 with table 1.

TABLE 4. PRESENT VALUE OF THE INCOME FROM AN ACRE OF DRAINED, TEMPORARY WETLAND (DITCH REGION)

	Interest rate (r)	
No. of years (n)	Competitive conditions (8½%)	Subsidized conditions (5%)
1		$28.38
2	$37.88	55.41
3	54.63	81.15
.	.	.
.	.	.
.	.	
7	109.48	
8	120.62	

The present value of the income stream exceeds the largest subsidized drainage cost (I_p) in the sample after three years, and capital is recouped on the average after two years.

Even under free market prices for agricultural outputs, commercial interest rates, and the landowner paying the full resource cost of drainage, the present value of the expected future income stream not only exceeds our estimate of the mean cost of draining temporary marshes, but also exceeds the drainage costs at the upper end of the distribution of cost outlays. This statement applies to both types of drainage systems, and is tantamount to contending that on temporary wetlands, where the land is cultivated despite inadequate drainage in the hopes of producing a weak crop, investment in drainage facilities is not only rational from a private point of view, but also socially desirable if the opportunity cost in terms of waterfowl is small enough. The statement is not, of course, an endorsement of public subsidies for such drainage.

Distribution of Temporary Wetlands When Additional Cultivation Costs Are Nonzero

So far the analysis has been restricted to the situation where occasional temporary wetlands dot an area that is predominantly well-drained. However, where the landscape is composed primarily of large complexes of poorly-drained soil, one must strain to make the argument that in the absence of adequate drainage the operator will still attempt to farm the land. My most optimistic estimates of predrainage revenues evaluated at support prices indicate that this land is marginal at best. For both southern and northwestern Minnesota (the tile and ditch drainage regions respectively) predrainage revenues barely exceed variable costs, thus making it questionable whether cultivation will be attempted even in the short run.[14] Hence, this case is unlike the case of isolated potholes, and cultivation costs will be incurred when drainage systems are installed on relatively large expanses of inadequately drained land. Although the criterion for determining the socially optimal allocation of the land is still 2.1, the discounted value of the income stream after drainage is altered to

$$V_{ap} = \sum_{t=1}^{n} \sum_{i=1}^{m} \frac{\beta_{is}(\overline{p_i q_i} - C_i) + C_d}{(1+r)^t}, \qquad (2.2)$$

where C_i is the total cost of cultivating $\overline{q_i}$, and the numerator is simply the net revenue from cropping an acre of land.[15]

In effect we are asking whether under free market conditions it remains profitable to drain and cultivate the land, whereas in

---

[14] My most optimistic estimates of the annual $E(TR - VC)$ were used in arriving at these conclusions. These estimates have a distinct upward bias, and are calculated on the basis of estimated per acre cost data supplied by the University of Minnesota Agricultural Extension Division and the previous estimates of gross, pre-drainage revenues (table 2A.3). As a result of the fact that the estimates are optimistic, the conclusion that cultivation in the absence of adequate drainage is unlikely is strengthened. (See the appendix to chapter 2 for the derivation and discussion of the data on the cost of cultivation.)

[15] Since we are dealing with newly cultivated land here, q_i (total output per acre) has been substituted for dq_i, and C_i (the total cost of q_i) has been substituted for dC_i. C_i includes an estimate of additional capital costs and land rent, additions which are legitimate because the investment under consideration is going to transform a marsh (whose rental value is assumed to be zero) into completely arable land. If one crops this land, rent is an appropriate expense.

the previous section we were concerned with whether drainage expenditures were justified given the assumption that an attempt to crop the land would be undertaken anyway.

The term C_d, the nuisance cost of wetlands, requires some discussion. Planting in irregular patterns increases cultivation costs. Machinery often becomes mired in the mud near a pothole, and it is expensive to extricate it. Waterfowl migrating south in the fall sometimes attack crops in the vicinity. The first two of these possible diseconomies are relevant for temporary wetlands, but there are no data available to estimate either the prevalence or the size of such diseconomies. It is not evident that such costs are even related to the size of the pothole (the variable of relevance here). Instead, these costs might be dependent upon the number of water areas or the perimeter or shape of the marsh. It is likely that the distribution of such costs would be widely dispersed, thus making inference of the mean to all drainage projects precarious. In any event, given the paucity of information about C_d, we are forced to disregard it, rationalize its lack of importance, and concede that our results will be biased in favor of retention of wetlands because it has not been incorporated into the model.

In determining the optimal allocation of wetlands the same technique was employed as was used for determining whether investment in drainage facilities was profitable under the assumption of zero additional cultivation costs. On the basis of alternative per acre cost data,[16] and our previous estimates of gross revenues ($\Sigma_{i=1}^{m}\beta_{is}\overline{p_i q_i}$ from table 2A.3), net revenue streams were calculated. In generating the income streams, alternative assumptions regarding crop distributions, estimates of output yields, and prices for the output were employed. Consequently, several possible net revenue streams were calculated, with each stream implying a different set of assumptions. The streams were classified according to whether they were evaluated by support or free market prices. The extreme values (lowest and highest values) of each group were discounted at a subsidized

---

[16] See appendix to chapter 2 for derivation and discussion.

TABLE 5. NUMBER OF YEARS BEFORE THE PRESENT VALUE OF THE INCOME STREAM IS IDENTICAL WITH THE INVESTMENT COST

Tile drainage

Estimate of net revenues	Full cost (I_s) Mean 5% 8½%	Full cost (I_s) Extreme 5% 8½%	Private cost (I_p) Mean 5% 8½%	Private cost (I_p) Extreme 5% 8½%
At support prices:				
Highest	6 6	8 10	3 4	5 6
Lowest	6 7	9 11	4 4	6 6
At free market prices:				
Highest	24 *	* *	11 14	21 *
Lowest	42 *	* *	15 25	35 *

Ditch drainage

Estimate of net revenues	Full cost (I_s) Mean 5% 8½%	Full cost (I_s) Extreme 5% 8½%	Private cost (I_p) Mean 5% 8½%	Private cost (I_p) Extreme 5% 8½%
At support prices:				
Highest	3 3	9 10	2 2	4 5
Lowest	4 4	12 17	2 3	5 6
At free market prices:				
Highest	11 16	* *	5 5	27 *
Lowest	* *	* *	100 *	* *

Notes:

1. Tile system assumed to depreciate completely in 20–50 years. Net income streams for ditch systems allow for maintenance, so ditch systems are assumed to generate revenue in perpetuity.

2. Highest and lowest estimates of net revenue streams reflect differential soil fertility and alternative assumptions about crop distributions. They do not reflect variations in estimated prices. A unique vector of free market and support prices was used. See table 2A.2.

* Investment is unwarranted. Equality will not occur within the productive life expectancy of the drainage system.

interest rate (5 percent) and at a rate taken as representative of the commercial one (8½ percent). The resulting present values were compared with the mean and extreme values for I_s and I_p. The results appear in table 5.

The entries in the table are the number of years necessary before the present value of the income stream from the reclaimed land becomes identical with the investment cost of drainage. An asterisk means that within the productive life expectancy of the drainage system this equality never occurs, i.e., the investment is unwarranted. It will be recalled that estimates of the time period over which a tile system completely depreciates range from twenty to fifty years. I assume that any tile project that takes more than twenty years for the present value of the income stream to cover the investment outlay is, at the very least, marginal. In computing the net income streams for the ditch systems, allowance was made for their maintenance, so such systems are assumed to generate revenue in perpetuity.

Analysis of Table 5

A number of fairly interesting conclusions emerge from table 5. It is evident that neither the ASCS's subsidy payments toward the cost of installing drainage systems nor the liberal loan policies of the FHA are responsible for the profitability of reclaiming and cultivating temporary wetlands. Even if the wetlands owner were forced to bear the full resource cost of a relatively expensive drainage system (extreme I_s) and to borrow at commercial rates (8½ percent), the present value of the income stream evaluated at support prices is still large enough to encourage the investment. This conclusion is applicable to both types of drainage and holds even for the lowest estimate of the net income stream. Under these assumptions the present value of the income stream and the investment cost are equal after eleven years for the tile system and seventeen years for ditch drainage. Thus all of the systems included in our sample would have been currently desirable investments from a private viewpoint in the absence of liberal loan policies and subsidies covering the initial installation cost.[17]

A markedly different set of conclusions ensues if free market prices for the agriculture sector are introduced. Consider tile-drained land first. With the removal of all three types of subsidies, drainage no longer proves desirable. If the wetlands owner were forced to bear the full resource cost of the drainage system, to borrow at commercial rates (8½ percent), and to receive free market prices for his produce, the investment outlay involved would exceed the discounted value of the revenue stream. Even the least costly system in our sample (not shown in the table) would have been unwarranted under the assumed conditions. Liberal (5 percent) loans to farmers would appear to make some of the average-cost projects marginal despite the free market prices, but none would become unusually attractive.

Similar conclusions follow for the ditch drainage systems. With low free market revenues and no installation subsidies, none of the drainage projects is justifiable. (Even low interest rates will not provide sufficient inducement for drainage.) The range of the per acre revenue estimates for the output of the land in northwestern Minnesota is such that the average drainage project is profitable if the highest free market revenues are used. However, relatively costly drainage systems are not attractive even when these extreme estimates for revenues are accepted.

Continued subsidization of the original investment cost of a drainage system (either tile or ditch) does make some projects privately profitable despite the imposition of free market agricultural prices. It is evident from the table, however, that this particular subsidy is not nearly as influential in inducing drain-

---

[17] This result (that with the existing support prices for commodities, the present value of the output exceeds the investment cost associated with the drainage of all temporary wetlands) is a bit unsettling. It makes it difficult to explain the continued existence of any privately-owned temporary wetlands. If it is currently profitable to reclaim them, why have they not all been drained? This point is discussed in the appendix to this chapter in the section "An Apparent Paradox in Table 5."

age as support prices are. Drainage installation subsidies alone are not sufficient to make the more costly projects attractive for either tile or ditch systems. The table shows that with free market prices and 8½ percent interest the extreme-cost projects are unwarranted even when revenues are high. For average-cost projects the situation is somewhat different. Because of the range of revenue estimates no universal conclusions are possible about the influence of installation subsidies on these projects. With an installation subsidy an average-cost tile project becomes profitable within twenty-five years if revenue is low and within fourteen years if revenue is high. This means that tile drainage, instead of being unwarranted, is marginal on the less productive land and worth considering on land that is unusually fertile and can produce high free market revenues. For the average-cost ditch drainage projects, the installation subsidies exert no influence if free market revenues are low, i.e., the project remains unwarranted. Where free market revenues are high, an installation subsidy results in the investment expense being recouped within five years instead of sixteen, and hence makes the project more attractive. It does not make the difference between profitability and nonprofitability, however. There must be some projects that generate revenues between these two extremes and that would therefore be marginal. Thus, no definitive result about the influence of installation subsidies emerges in the case of projects of average drainage expense whose output is marketed under free market conditions.

On the basis of our sample of drainage projects, the drainage of large complexes of temporary marshes in southern Minnesota (the tile region) would be socially undesirable under competitive conditions. Since current market clearing prices would be less than free market prices for agricultural produce, the commitment of additional resources for drainage and cultivation constitutes a present misallocation of resources. At the very least, those projects of above-average expense in northwestern and west-central Minnesota (the ditch drainage area) can be similarly criticized. Due to the range of estimated values for free market prices no unambiguous conclusion emerges regarding the remaining ditch drainage projects. Some might be warranted, but they are unlikely to be very profitable.

It should be emphasized that these results were derived under the assumption that these marshes have no social value in the wildlife sector. To the extent that large expanses of temporary wetlands contribute to waterfowl production, the opportunity cost of reclaiming land is increased, and the case against drainage is strengthened. It is very likely, however, that as long as these wetlands are isolated from more permanent habitat, the assumption of zero wildlife value is correct.[18]

The Allocation of Permanent Wetlands

The analysis thus far has been restricted to temporary wetlands. Although direct subsidies for drainage of permanent water areas are no longer available, concern regarding their possible conversion to arable land continues to be voiced by waterfowl enthusiasts.

Permanent wetlands, unlike temporary ones, have a marketable product because they provide opportunities for recreational hunting. Make the following assumptions: (1) permanent wetlands in their natural state can be utilized only as recreational waterfowl hunting sites, (2) hunting rights are sold under competitive conditions,[19] and (3) competitive conditions exist in the agriculture sector. (The consequences of relaxing these unrealistic assumptions will be considered later.) Under these conditions there is no divergence between private and social values for this resource, and its market allocation will be optimal. If hunters are willing to pay sufficiently high rental prices, permanent water areas will be retained in the wildlife sector. Alternatively, if agricultural land becomes sufficiently scarce, wetlands

---

[18] See chapter 4, pp. 67–68.

[19] Under current conditions, of course, privately owned hunting land must compete with entry-free, public hunting land. Although hunters may relish the exclusive rights to hunting land and hence may be willing to pay a positive price to use it even when public land has a zero price, it is clear that the price commanded by private hunting land under these conditions is less than it would be if the rights to all hunting land were openly marketed. The implications for optimality which this mix of public and private ownership poses are discussed in chapter 5.

will be reclaimed. The condition for allocation at the margin is the same as that previously specified in 1.4, page 6.

if $V_{ws} > V_{as} - I_s$, the land will remain a marsh;
if $V_{ws} < V_{as} - I_s$, the land will be drained.

Before proceeding with the attempt to estimate $(V_{as} - I_s)$ for permanent wetlands, it seems appropriate to elaborate a little on expression 1.2, which is the expression for the social value of wetlands (V_{ws}). We have retained the assumption introduced at the beginning of this chapter that permanent wetlands serve only as a landing field for ducks, i.e., that $W = 0$. To the extent that permanent wetlands are also an essential part of the breeding ecology of waterfowl, there will be a social value that is not accounted for in the market. The problem of quantifying this value is discussed in a later chapter. The trapping rights for aquatic furbearers may be another significant revenue source which is precluded by assumption above. Evidently, sale of such rights is common, but estimation of their value was considered to be beyond the scope of this study. It is important to recognize that V_{ws} will be underestimated if these values are neglected.

Each of the two cost terms in V_{ws} has been assumed equal to zero. The maintenance cost of operating a hunting slough (C_w), is very likely negligible, consisting primarily of occasional advertising expenses. Some of the nuisance costs (C_d) resulting from the presence of wetlands in the midst of cultivated land were discussed earlier, but not the ravaging of crops by migratory waterfowl. Grain-feeding ducks attracted by a marsh represent a potential threat to the crops in the vicinity. Several studies have attempted to estimate the expected cost of waterfowl-inflicted damage, but the estimates exhibit wide variation, and the authors regard their results as inconclusive.[20] Statistically

[20] Gene Lee Wunderlich, "Private Costs from Public Benefit Projects: A Case Study of Waterfowl Depredations on Small Grains in North Dakota," Master's thesis, Iowa State University, 1951. Merrill C. Hammond, "Ducks, Grain, and American Farmers," *Waterfowl Tomorrow*, U. S. Department of the Interior, 1964, p. 423. Ernest L. Paynter and W. J. D. Stephen, "Waterfowl in the Canadian Breadbasket," *Waterfowl Tomorrow*, p. 409–416.

the average damage inflicted upon individual farms is not significantly different from zero, but depredation costs have been reported which run as high as $44 a day per 1,000 ducks attracted to the area.[21] The amount of damage inflicted is much more a function of the weather than of the size of the fall flight. If the weather is dry, much of the grain is harvested prior to the fall migration. Even if it is not fully harvested, but merely swathed and lying in the fields, the birds are not particularly attracted to it.[22] The only statistically significant results stemming from the research on this problem are that damage inflicted is a decreasing function of the distance between the grain field and the birds' resting area,[23] and that the problem seems more severe in Canada than in the United States.

Due to the inconclusive nature of these studies, we shall assume that such expected costs are zero. This is not an unrealistic assumption in light of the evidence that the distribution of ravaging costs per acre of wetland is skewed markedly to the right with most of the values close to zero. Nevertheless, it should be recognized that this assumption will bias our conclusions in favor of the retention of wetlands. There are two situations, however, in which this bias will not weaken our conclusion about the distribution of permanent wetlands. (1) If it is found that reclamation of permanent wetlands is warranted, then our position is quite strong, for this result will have been reached despite the fact that depredation costs, costs which would have provided additional incentive for drainage, were

---

[21] Hammond, "Ducks, Grain, and American Farmers," p. 423.

[22] Al Studeholm, Fish and Wildlife Service, U.S. Department of the Interior, Washington, D.C.

[23] If crop ravaging is not restricted to the property of the owner of the hunting slough, our contention above that there is no divergence between private and social values for permanent wetlands and that their market allocation will be optimal will no longer be true. Damage inflicted on individual A's crops by the ducks attracted to neighbor B's marsh constitutes an external diseconomy. A bears the cost, and B receives the benefits. From a social point of view, such an externality would not alter the direction of the inequality in expression 1.4. But private and social values would differ, and thus 1.4 would not be the criterion used by participants in an unregulated market to distribute wetlands. Too many wetlands would exist if this externality were not corrected.

assumed zero. (2) If it is found that drainage of permanent wetlands would only be justified if crop ravaging were severe, then the assumption again does not temper our result, for the evidence indicates that on the average such damage is small. Only at the margin where a relatively small amount of depredation would make the difference between the land being preserved as wetland and being drained will it be difficult to make a conclusive statement regarding the proper allocation of the land. All of these arguments apply also to the other components of C_d, the nuisance cost of wetlands, components which were also assumed to be zero.[24]

Statistical Results

As was noted in the introduction to this chapter, standing water may result (1) from an impermeable subsoil which forms a pothole through which water drains very slowly or (2) from a high water table. If the water table is high, both an outlet ditch and tile mains will be required to drain the area, the ditch being installed first to drain off the surface water and lower the water table sufficiently to permit tile placement. The ditch outlet must be dug in stages, because the equipment is heavy and cannot penetrate very far into the marsh while the land is wet. In the case of a pothole, a ditch may be sufficient for permanent drainage if the impermeable soil layer lies close to the surface. We shall assume that this is not the case, that in draining most permanent water areas a tile system will be needed in conjunction with the ditch outlet.

Under the assumption that drainage installations are approximately a third more expensive per acre[25] for permanent water

---

[24] See p. 27.
[25] No sample of permanent water area drainage projects was taken. Such projects are ineligible for governmental subsidies under the law, and hence no data are collected by public agencies regarding such projects. No estimates are available of the number of acres of permanent wetlands which are drained annually. Nor is there any accurate and accessible source for cost data associated with such drainage. A pilot sample of private drainage contractors was instituted in order to determine whether their records could be used to estimate per acre costs. It proved abortive (1) because of an inability to distinguish temporary from permanent wetlands on the

areas, the following cost estimates were obtained: the mean I_s is $209.98 per acre,[26] and costs range from $164.72 to $304.54, I_s being the full resource cost of draining an acre.

Since new land is being brought into cultivation, additional costs will be incurred in growing the output. Hence, the revenue estimates net of production costs are the relevant ones to use in computing a present value for the income stream of this land. Table 6 gives the present value of an acre of reclaimed

TABLE 6. PRESENT VALUE OF AN ACRE OF RECLAIMED LAND EVALUATED AT FREE MARKET PRICES WITH A DISCOUNT RATE OF 8½ PERCENT

Net revenue at free market prices	Southern Minnesota		Northwestern Minnesota	
	20 years	50 years	20 years	50 years
Highest	$109.77	$134.16	$71.35	$87.20
Lowest	86.12	105.25	26.02	31.80

land (V_{as}) for alternative time periods, using free market prices and a discount rate of 8½ percent. It is quite evident that even the most liberal estimate ($134) falls far short of covering even the smallest element in our sample of drainage costs ($164). The magnitude of the difference between V_{as} and I_s is such that the nuisance cost of wetlands would have to be much larger than is indicated by any of the evidence in order to reverse the result that reclamation of permanent marshes is not warranted if perfectly competitive prices prevail in the agriculture sector.[27]

---

basis of such records, and (2) because of cavalier acreage estimates by the contractors.

The per acre cost estimates used here are based entirely on the opinions of experts in drainage technology. It was the general, though not uniform, consensus of these experts that drainage of permanent wetlands necessitated both a ditch and a tile system, and that the tile system would increase drainage cost by approximately one-third. The individuals consulted were Evan R. Allred and Philip W. Manson, Department of Agricultural Engineering, University of Minnesota, Ross St. John, chief engineer, SCS, St. Paul, and other members of the SCS in Minnesota.

[26] This is 4/3 times $157.49. See table 1.

[27] For southern Minnesota the annual nuisance cost of an acre of wetland (C_d) would have to be sufficient to increase the discounted value of the income stream after drainage by about $100 (or more than $10 a year) to even qualify the project as marginal. For northwestern Minnesota the discounted C_d would have to be about $140. The available evidence indicates that nuisance costs per acre are not likely to be this large.

Private Profitability of Reclaiming Permanent Wetlands under Conditions of Price Supports

Since it has just been shown that drainage of permanent wetland is socially undesirable, it is quite important to determine whether current (1963) support prices for agricultural produce are sufficient to induce such drainage.[28] Although a uniform cost of drainage has been assumed for permanent water areas regardless of their location, the annual net revenue is dependent upon the productivity of the soil, and hence is not independent of location. Using the previous estimates of annual revenue per acre net of production costs, and discounting at a rate of 8½ percent for twenty years and fifty years, we obtain the estimated present values shown in table 7. The rationale for the twenty-

TABLE 7. PRESENT VALUE OF AN ACRE OF RECLAIMED LAND EVALUATED AT SUPPORT PRICES WITH A DISCOUNT RATE OF 8½ PERCENT

Net revenue at support prices	Southern Minnesota		Northwestern Minnesota	
	20 years	50 years	20 years	50 years
Highest	$342.85	$419.02	$193.62	$236.63
Lowest	316.64	386.98	151.41	185.05

and fifty-year periods is the same as before, i.e., these are the extreme values for the time period over which the system is expected to depreciate.

Temporarily make the assumption that the rental value of recreational hunting land is zero, and within this assumption analyze the distribution of permanent wetlands. This restrictive assumption will be dropped in chapter 5, and the results modified as necessary at that time.

It is clear from table 7 that support prices make it profitable to reclaim the relatively fertile lands of southern Minnesota. Even the highest-cost drainage (I_s) in our sample ($304.54) is less than the present value of the reclaimed land when price

---

[28] Although 1963 price supports were used in the analysis, the revenue estimates are likely to be still valid. Wheat was the only commodity whose support price in 1970 was significantly different from that in 1963, having fallen 30 percent. However, wheat represents a small proportion of the output in the region being considered.

supports are in effect. This does not mean that all the waterfowl habitat in southern Minnesota is in jeopardy, but it does constitute strong evidence that drainage of at least some of it may be induced by price supports. Any nuisance costs that might emanate from wetlands would strengthen this conclusion, for they would increase the incentive for drainage.

For northwestern Minnesota, where the land is not nearly so productive, a drainage project that involved average investment expenses ($209.98) would be marginal at best. On the basis of our cost and revenue estimates and the assumption that $C_d = 0$, such a drainage system would have to endure for at least twenty-six years and its output would have to be valued at our highest estimate of support revenues for its agricultural value (V_{ap}) to equal its drainage cost (I_s). Thus, drainage of permanent water in the northwest region of Minnesota is not very likely. However, if it proves possible to drain permanent water in the northwest with a ditch outlet alone, or if annual nuisance costs are moderate ($2 to $7 per year), then such drainage will be more extensive than the above data suggest.

Summary

Despite all the restrictive assumptions that had to be made throughout this section of the analysis, it was possible to derive some important conclusions about the allocation of wetlands. These conclusions are summarized here, but it should be emphasized that the results were derived without reference to any value that wetlands might have in the wildlife sector. To the extent that wetlands are productive in their natural state, the opportunity cost of reclaiming land is increased, and the case against drainage is strengthened. Later in this paper the rental value of wetlands for hunting purposes (R) is investigated. A finding that rental values are high enough to forestall drainage would alter the conclusion that permanent wetlands in Southern Minnesota can be drained profitably if current price supports are maintained. None of the other conclusions listed below would be affected.

Temporary Wetlands

(a) *Zero additional cultivation costs.* Under the assumption that an attempt will be made to farm the land in any event, it is profitable to engage in tile and ditch drainage. This is true even when the owner of temporary wetlands must bear the burden of unsubsidized conditions in all markets, i.e., competitive prices for agricultural produce, commercial interest rates, and the full investment cost of drainage. Subsidies granted by the USDA did not make the difference between profitability and nonprofitability of draining any of the temporary wetlands in the sample. This is not to say that these subsidies did not induce more wetlands to be reclaimed than would have been drained in their absence. The subsidies did alter the profitability of drainage relative to all other possible investments. However, given normal returns on capital throughout the rest of the economy and the assumption that temporary wetlands have no alternative use in the wildlife sector, then the subsidies have not seriously affected the distribution of resources.

(b) *Nonzero cultivation costs.* (i) Southern Minnesota (tile): If all of the subsidies are removed, none of the projects are warranted. Support prices for farm output alone, however, are sufficient to make even the most expensive drainage project in the sample profitable. (ii) Northwestern Minnesota (ditch drainage): Some of these drainage projects may be economically feasible in the absence of all subsidies. But facilities that involve relatively high, initial investments are not warranted in this area. The granting of price supports for farm output, however, is sufficient to induce the undertaking of even the more costly projects. (iii) Support prices for agricultural output are more influential than the subsidies to capital expenditure in inducing drainage of temporary wetlands.

Permanent Wetlands

(a) *Perfectly competitive farm prices.* The investment cost involved in reclaiming permanent wetlands precludes drainage if perfectly competitive prices prevail in the agriculture sector.

(b) *Support prices for farm products.* (i) Southern Minnesota: The fertility of the land in this region is such that drainage is profitable under current price subsidies. (ii) Northwestern Minnesota: The land in this region is not nearly so productive as land in the south, and hence a drainage facility which involves even average investment expense is marginal at best.

Appendix to Chapter 2

TABLE 2A.1. CHANGES IN YIELDS PER ACRE RESULTING FROM DRAINAGE

	Tile drainage area (Southwestern and Southern Minnesota)		Ditch drainage area (Northwestern and West-central Minnesota)	
Crop	$\overline{dq_i}$	$\overline{q_i}$	$\overline{dq_i}$	$\overline{q_i}$
Corn (bu.)	53.4	95.2	43.0	65.8
Soybeans (bu.)	23.2	37.1	17.7	25.7
Wheat (bu.)	26.6	47.4	15.0	25.8
Oats (bu.)	47.7	85.2	38.8	62.4
Barley (bu.)	—	—	20.3	36.9
Hay (ton)	2.8	4.8	2.1	3.2

TABLE 2A.2. COMPARISON OF ESTIMATED FREE MARKET PRICES AND 1963 SUPPORT PRICES FOR SELECTED CROPS

		1963 support prices[b]	
Crop	Free market prices[a]	Tile drainage area	Ditch drainage area
Wheat (bu.)	$ 0.87	$ 1.97	$ 1.97
Oats (bu.)	0.41	0.62	0.58
Corn (bu.)	0.77	0.98	0.96
Soybeans (bu.)	1.35	2.19	2.15
Hay (ton)	15.40	19.60	19.20
Barley (bu.)	0.62	—	0.82

[a] Estimates of free market prices are from *Economic Policies for Agriculture in the 1960's*, Joint Economic Committee (Washington, D.C.: Government Printing Office, December 1960), p. 17. There are a number of other studies with similar price projections based on slightly different assumptions. The differences in their predictions are small. This particular study was chosen because it was the only one with price predictions for all the commodities I was concerned with, except hay. A ton of hay, which has approximately twenty times the feed nutrient value of a bushel of corn, was assigned a price equal to twenty times the price of corn (Source: Dean Campbell, SCS, Mankato, Minnesota).
[b] Support prices are taken from *1963 Crop Loan and Purchase Agreement Program*, USDA. Reprinted from *Federal Register*, 1963, title 7, parts 1421.2113, 1421.2609, 1421.2310, 1421.2910, and 1421.2210.

TABLE 2A.3. RANGE OF VALUES FOR ANNUAL ADDITION TO GROSS REVENUES FROM AN ACRE OF DRAINED WETLAND

	Tile drainage area (Southwestern and Southern Minnesota)		Ditch drainage area (Northwestern and West-central Minnesota)	
	Support prices	Free market prices	Support prices	Free market prices
$\sum_i p_i \beta_i \overline{dq_i}$	$47.11–49.52	$33.01–34.70	$29.80–35.06	$21.39–26.72
$\sum_i p_i \beta_i \overline{q_i}$	83.75–85.31	58.18–61.04	47.14–54.16	33.89–41.24

41

Per Acre Cost of Cultivation by Crop

A number of studies estimating the per acre cultivation cost of crops by region in Minnesota are available, but only those issued by the University of Minnesota Agricultural Extension Service incorporate an estimate of depreciation on equipment, rental value of land, and interest on capital invested in machinery. Since depreciation on farm equipment seems to stem directly from additional use and not from obsolescence, such expenses need not be classified as joint and unassignable, but can legitimately be considered as increasing when new acreage is brought into production. Included in the cost estimates given below is a $1.25 per hour charge for labor. Since the Agricultural Extension Service studies did not include cost estimates for all of the crops which we were interested in, an average cost for the excluded commodities was arbitrarily assigned, and was derived from other available studies.[29]

TABLE 2A.4. PER ACRE CULTIVATION COSTS BY CROP (C_i)

Crop	Southern Minnesota	Northwestern Minnesota
Corn	$56.62	$38.65
Soybeans	42.25	28.80
Wheat	33.00	33.75
Oats	48.80	28.75
Hay	55.95	30.00
Barley	—	30.75

Sources: For Southern Minnesota: Hal Routhe and Duane Erickson, *Which Are Your High Return Crops?* University of Minnesota Agricultural Extension Service, December 1962. For Northwestern Minnesota: Paul Hasbargen, *Which Are Your High Return Crops?* University of Minnesota Agricultural Extension Service, January 1963.

---

[29] The two excluded commodities were wheat in the south and hay in the northwest. The sources from which a cost estimate for each of these crops was derived were: (1) Donald Taylor, Ph.D. thesis, Department of Agricultural Economics, University of Minnesota, St. Paul, 1965, pp. 51–52; (2) W. B. Sundquist, L. M. Day, and H. R. Jensen, *Profitable Adjustments in Farming in Central Minnesota*, University of Minnesota Agricultural Experiment Station, Bulletin No. 460, April 1962, pp. 11–12; (3) C. O. Nohre and H. R. Jensen, *Profitable Farm Adjustments in South-Central Minnesota*, University of Minnesota Agricultural Experiment Station, Bulletin No. 471, 1964, p. 9.

The cultivation costs in table 2A.4 together with the alternative crop distributions given in table 2 were used to derive the per acre cost figures in table 2A.5. The calculation performed and the entries in table 2A.5 are simply $\sum_{i=1}^{m} \beta_{is} C_i$. The assumption implicit in this table is, of course, that these per acre costs apply regardless of yield.

TABLE 2A.5. UNIFORM COST OF CULTIVATING AN ACRE OF OUTPUT ($\sum_{i=1}^{m} \beta_{is} C_i$)

	Tile			Ditch		
	β_{i1}	β_{i2}	β_{i3}	β_{i1}	β_{i2}	β_{i3}
$\sum \beta_{is} C_i$	49.08	48.20	50.29	33.70	33.07	31.14

An Apparent Paradox in Table 5

The data indicate that with the existing support prices for commodities, the present value of the output exceeds the investment cost associated with the drainage of all temporary wetlands. In fact, this is the case even if the wetlands owner were denied direct subsidization of the installation cost and low rates of interest.[30] This result makes it difficult to explain the continued existence of any privately-owned temporary wetlands. If it is profitable to reclaim them, why have they not all been drained?

There are several possible explanations. Due to the first-come-first-served distribution of subsidies by the ASCS, eligible wetlands are often denied grants. Although it may prove profitable to finance the entire drainage system privately, the prospect of future government assistance may induce a farmer to postpone drainage. Such postponement would not be indefinite, however, and the long-standing nature of the subsidy programs would seem to temper the force of this argument.

Secondly, although there may be only two conceivable uses for wetlands, there are a number of alternative opportunities for financial capital. Capital constraints result in only the most

[30] It is entirely possible for this denial to occur despite the fact that the wetland in question is legally eligible for these subsidies. The ASCS is restricted by a limited Congressional allotment. Normally, applications for assistance more than exhaust the available funds.

profitable projects being undertaken. Consequently, more pressing uses for capital may cause the deferment of drainage, and account for the phenomenon of undrained lands.

Finally, it is possible that a bias has been introduced into the sample which results in an overestimate of the profitability of reclamation.[31] The sample used in estimating the cost of draining an acre of land was drawn not from the population of all temporary wetlands but from the population of wetlands that had been drained. Since these lands were drained because someone judged it profitable to do so, it follows that $V_{ap} - I_s$ is positive for all projects in this group.[32] If such an upward bias has been introduced, the inference that drainage of large complexes of temporary wetlands is privately profitable given support prices for agricultural produce is warranted only for those projects whose capital cost is similar to that of the projects currently being assisted by SCS.

The existence of some temporary wetlands under current conditions is not an enigma, then. Any one of the above three explanations provides a plausible rationale for their continued presence.

[31] Anne Krueger, Department of Economics, University of Minnesota, suggested the explanation of a sampling bias.
[32] V_{ap} evaluated at support prices.

3

A MODEL FOR EVALUATING WATERFOWL AND THEIR ENVIRONMENT

Under the simplifying assumption that waterfowl have no value, it was possible to derive a number of significant conclusions about the allocation of wetlands. That assumption is abandoned in this chapter, and an attempt is made to evaluate waterfowl. However, no attempt has been made to incorporate into this analysis either the benefits that accrue to bird watchers or the vicarious satisfaction enjoyed by those who derive utility from the knowledge that waterfowl exist. It is assumed that waterfowl have value only for duck hunters. Although some types of waterfowl are more highly prized as trophies than others, it was considered impractical to attempt to differentiate between species.

It seems appropriate to state at the outset that no satisfactory estimate of the social value of waterfowl or their habitat was secured. While the outcome is not a fortunate one, it is not entirely surprising because attempts at estimating nonmarket values are often not crowned with ultimate and complete success. Research in this realm is in its incipient stages, and the development of applicable techniques proceeds slowly.

The contribution of this research lies in the formulation of a framework within which the problem of pricing waterfowl (and perhaps other nonmarket commodities) may be appropriately studied.[1] Several models were considered. This chapter

[1] I am indebted to O. H. Brownlee, Department of Economics, University of Minnesota, Allen V. Kneese and John Krutilla of Resources for

is devoted primarily to a discussion of that particular model which seems to hold the most promise for future success in evaluating waterfowl and their habitat.

Invalid Evaluation Procedures

Proposals for estimating the value of nonpriced commodities run the gamut from the incredibly naive and invalid to the more conservative attempts to apply conventional analysis.[2] The most commonly encountered error involves double counting in one form or another. Very often analysts will engage in an enumeration of all the expenditures incurred by a participant in an activity, inferring that these total expenditures are directly assignable as the "value" of the activity to the individual. For example, one might propose aggregating the expenditures incurred by duck hunters (ammunition, depreciation on their guns, lodging, food expenses over and above those which would have been incurred at home, gasoline, income forgone, etc.), dividing the total by the number of ducks bagged, and calling the resulting statistic the value of a duck. Clearly, the statistic resulting from this manipulation is not the value of a duck. What hunters have purchased with these expenditures is not ducks, but the recreational experience of duck hunting. Even as a gross measure of the value of the recreational experience these aggregate expenditures are incomplete, because they fail to include the cost of the resources contributed by public

---

the Future, and to Charles Howe, Andrew Sheldon, and Robert Steinberg, formerly at Resources for the Future, for their suggestions and assistance on the methodology. The responsibility for the results is, of course, mine alone.

[2] For a partial list of both valid and invalid techniques see James A. Crutchfield, "Valuation of Fishing Resources," *Land Economics*, May 1962, pp. 145–54; and John B. Moyle, *Review of Approaches and Methods for Estimating Values of Fish and Game and of Hunting and Fishing*, Special Publication No. 14, State of Minnesota Department of Conservation, St. Paul, Minnesota. Incisive discussions also appear in Outdoor Recreation Resources Review Commission Report 3 (*Wilderness and Recreation—A Report on Resources, Values, and Problems*) and Report 24 (*Economic Studies of Outdoor Recreation*), Washington, D.C., 1962.

agencies, and, except in rare instances, the rental value of the recreational site.[3]

Even if one were willing to concede that the recreational experience itself provides no satisfaction to hunters, that the only thing hunters crave is birds in the bag, the averaging technique described above would still not qualify as an appropriate procedure for evaluating waterfowl. The relevant measure of the value of waterfowl is the amount that hunters are willing to pay at the margin in order to acquire one more bird, that is, the amount that they are willing to pay after having incurred all the other expenses associated with the hunting experience.[4] These other expenses are no direct indication of the price of the waterfowl themselves.

Fee Hunting: A Possible Market Value for Waterfowl

Although most hunting does not occur under circumstances where a direct fee is imposed for each bird bagged, private game farms exist which do exactly this. Birds are released for customers to shoot at in a simulated wilderness environment.

---

[3] A more extensive treatment of "rental value" appears in chapter 5, but it seems appropriate to note here that the rental value of the recreational site is the *net* value of the recreational experience. If one aggregates the total expenditures which an individual incurs as a result of participating in an activity, one has included the cost (value) of many intermediate goods, and cannot represent this total figure as the net value of the recreational experience. Hence, the use of the term "gross" above. Perhaps an analogy will prove useful. Consider an individual who attends a movie. His total expenditures for this experience are the cost of a baby sitter, transportation expenses, parking fees, and the price of the ticket of admission. The net value or price of the movie is only the cost of admission to the theatre, not the total expenditures for the entire evening. Similarly, it is the rental price of the waterfowl hunting slough (the price of admission) that measures the net value of the activity itself.

[4] It is important to realize that the value of the last or marginal bird is less than that of the intramarginal birds. If one were concerned with the opportunity cost in terms of lost waterfowl of draining only one acre of wetland, a lower price would be used to value the birds than would be used to evaluate the birds lost as a result of reclaiming 100,000 acres of habitat.

Hunters are charged for each bird bagged, with prices averaging about $5.00 per duck.[5]

Given the synthetic character of the hunting experience on such game farms, one might think that $5.00 could be accepted as a lower bound for the value of a duck.[6] This is not a justifiable assumption. The appeal of most of these private game preserves lies in their location close to centers of population. In a study conducted by Olin Mathieson,[7] successful preserves were an average of only 28 miles from population centers of 25,000 or more, while unsuccessful preserves averaged a distance of 37 miles from such population centers. The study also revealed the expected result that preserves close to population centers received considerably more patronage than those farther away. It might be valid to accept $5.00 as a minimum price for a duck if all natural hunting areas lay close to hunters' residences, but this is not the case. In a survey of duck hunters which I took with the help of the Department of the Interior[8] the average distance traveled by waterfowl hunters was 70 miles. One of the implications of the Olin Mathieson study is that patronage for fee-charging game preserves falls off severely when the average distance from a population center is increased from 25 to 37 miles. If this result is correct, $5.00 is likely to be an overestimate of the minimum price that hunters would be willing to pay for a duck in a natural setting about 70 miles from their homes. Since the suitability of such a fee as a lower bound for the value of a duck seems so suspect, and since most hunters consider the experience at game farms to be a highly imperfect substitute for hunting in the wild, I do not think that the fee

---

[5] *1963–64 National Shooting Preserve Directory*, Sportmen's Service Bureau, New York, N.Y., and *Economic Survey of Shooting Preserves*, Olin Mathieson Chemical Corp., East Alton, Illinois, 1959.

[6] Some writers have suggested as an appropriate lower bound the price charged in supermarkets for game. However, I think that the two commodities are not comparable, and it would be a severe error to assume that the duck hunter's purpose is to reap the food value of the bird. See Crutchfield, "Valuation of Fishing Resources," p. 150, for additional discussion.

[7] *Economic Survey of Shooting Preserves*, p. 4.

[8] This survey is described and the results analyzed later in the chapter.

charged at such farms is a reliable measure of the value of a bird.

A Suggested Model for Evaluating Waterfowl and Wetlands

Since the results of the examination of fee hunting were not encouraging, several other approaches were investigated. None was completely successful, but the one discussed below seems to hold promise for providing insight into the value that hunters place upon waterfowl and their environment, even though attempts at estimating the parameters in the model met with frustration, the statistical fits being exceptionally poor. Undoubtedly the model needs refinement, and the methods for generating the data have to be improved. Nonetheless the structure of this model remains inherently appealing, for it resolves many of the methodological problems encountered in attempting to estimate the value of waterfowl, their breeding grounds, and recreational hunting land, and it employs the observable behavior of hunters in obtaining these estimated values. Although the model is applied here only to duck hunting, it clearly can be modified for use in other recreational pursuits.

The order employed in the exposition is to develop a procedure for estimating the value that hunters assign to the experience of recreational hunting, and, consequently, to the wetland resources as both hunting and breeding habitat; and then to derive from this the value of waterfowl and, subsequently, the value of breeding habitat alone.

The data to estimate the parameters in the model were collected in a series of mail sample surveys (see the appendix to this chapter).

The procedure for estimating the value of wetlands is an adaptation of a methodology originally suggested by Harold Hotelling for estimating the value of national parks and other recreational areas.[9] There are 87 counties in Minnesota, and

---

[9] The seminal work in this realm was done by Harold Hotelling (in Roy A. Prewitt, *The Economics of Public Recreation—An Economic Study of the Monetary Evaluation of Recreation in the National Parks*, National Forest Service, Washington, D.C., 1949). Hotelling's idea has

the model as initially conceived estimated one equation for each county. The equation for county i $(i = 1, \ldots, 87)$ is

$$DH = f(C, Y, Q, G), \qquad (3.1)$$

where

DH = the days hunted by an individual hunter in county i during an entire hunting season,
C = the round trip travel cost from the hunter's county of origin (his home county) to county i (the county in which he does his hunting),
Y = family income,
Q = a measure of the hunting quality in county i,
G = a gravity variable which measures the quality of the hunting near the hunter's home county.

One would expect the coefficients of C and G to be negative and those of Y and Q to be positive.

The character of the data to which this model was fitted is described in detail in the appendix to this chapter, but some explanation here of the data used for variables C, Q, and G may facilitate the exposition. In assigning a value to variable C, it was assumed that the individual always hunts in the same county. Clearly, hunters are not always so habitual, and may visit several counties during the season. Designing the model and the sample questionnaire becomes considerably more difficult in the absence of this assumption, however. The sample participants were asked in what county they hunted most frequently, and it was that county which the respondents were assigned as their county of destination. The distance from the home county to the destination county was measured by the distance between the population centers of the two counties. Such data are not always readily available, but a study estimating these distances had recently been done by John Borchert, Department of Geography, University of Minnesota, Minne-

---

been ingeniously developed and elaborated upon by Marion Clawson, *Methods of Measuring the Demand for and Value of Outdoor Recreation*, Reprint No. 10, Resources for the Future, Inc., Washington, D.C., February 1959; and Jack L. Knetsch, "Outdoor Recreation Demands and Benefits," *Land Economics*, November 1963.

apolis. If a respondent hunted within his home county, he was assigned a number judged to be the average distance traveled within that county. The cost of travel may be evaluated using *Cost of Operating an Automobile* (U.S. Department of Transportation, Federal Highway Administration, Washington, D.C., February 1970).

Several measures of quality are available, but the one ultimately chosen was derived from data collected by the Minnesota Department of Conservation. This department conducts a sample of all hunters in Minnesota who buy a small game hunting license. Hunters are asked how many ducks they shot during the season, in what counties the birds were shot, and how many days they participated in hunting in each county. Although experience indicates that only ten percent of the license purchasers respond to the sample, one is still able to derive from this data an estimate of the average number of ducks shot per day per hunter for a given season for each county in the state. I do not know whether such elaborate data are available for regions other than Minnesota.

The same data were used to estimate the gravity variable, G. Several alternative forms were constructed, an example of which is $\Sigma TK_j/d_j$, where TK_j is the total number of waterfowl killed in county j during the hunting season in question, d_j is the distance from county k (the hunter's home county) to county j, and the summation is over all j counties within 50 miles of county k.

Fitting a least-squares regression for each of the 87 counties is statistically impossible, because for each of the i counties Q_i is a constant. A coefficient cannot be estimated unless Q_i has some variability. Accordingly, the state was partitioned into five hunting regions, and equation 3.1 was estimated for each of these regions.

The next step is to evaluate each of the five regional equations by setting C, Y, Q, and G equal to their mean values for each region, i.e., each of these variables will have a mean value for Region 1, one for Region 2, and so on. This procedure estimates the mean value of DH for each of the five regions. This mean value is interpreted as the average number of days of hunting desired by a hunter who hunted in region m ($m = 1, \ldots, 5$). Assume that the percentage of the entire duck hunting population of Minnesota who hunted in region m is identical with the percentage of the sample respondents who hunted in region m. Since an estimate of the number of active hunters in Minnesota (about 110,000 in 1963) is available, it is possible to estimate the number of active hunters in each region, and consequently the number of hunter days demanded in each region. Adding the days hunted regionally gives the total number of days of hunting that hunters will engage in when no entry fee is charged. This is a point on the aggregate demand curve for hunter days in Minnesota. Hence, the five estimated equations may be viewed as the relations which generate the aggregate demand curve for hunter days in Minnesota.

The procedure for deriving the remaining points on the demand curve is quite straight forward. Introduce a hypothetical entry fee, f, and assume that hunters will respond to the imposition of an entry fee in exactly the same manner as they would respond to an increase in C, their round trip travel cost. To each mean value for C in the five regional equations for DH add f, compute the new DH, multiply by the appropriate population correction factor, and aggregate over the five regions to find the new total number of hunter days desired. This provides a second point on the demand curve for hunter days, i.e., the total number of days desired when entry fee f is charged. The remaining points on the demand curve are generated by choosing successively higher values for f, and repeating this general procedure.

It is very likely that this demand curve systematically underestimates the number of hunter days demanded at each and every value of f because it was assumed that hunters would respond to an increase in the entry fee (f) in exactly the same manner as they would react to an increase in travel costs (C). It is really inappropriate to attribute all of the reduction in recreational hunting participation that occurs when C increases to the higher monetary costs (C) alone; longer traveling times also discourage participation. Since the time cost associated with hunting in a given region was not increased when an entry fee was imposed, the assumption that hunters react to an increase in f exactly as they would to an increase in C is not entirely true, and the true reduction in hunter days demanded is probably less than our estimated relation suggests. This is not necessarily an undesirable feature of the analytic technique, as will be seen later, for it results in a conservative estimate for the value of wetlands.

In order to use this demand relation to determine the value of wetlands, an assumption must be made about the marginal cost of maintaining marshes and of breeding waterfowl, and hence about the marginal cost associated with producing alternative levels of hunter days. If this cost is assumed to be zero, then the area under the demand curve is the annual consumer surplus from recreational hunting when $f = 0$. Capitalizing this value at some appropriately chosen discount rate gives us the value of the marshlands (both breeding and hunting land) responsible for generating this hunting experience. However, the marginal cost of producing hunter days is very likely not to be zero. At least some portion of the costs imposed upon farmers who have wetlands on or near their property should legitimately be charged against the production of hunter days. In addition, in chapter 5 it is argued that hunters impose costs upon each other when they crowd onto marshes. Consequently, the optimal fee probably should be positive. The imposition of such a fee does not alter the procedure for evaluating wetlands, however. It merely reduces the size of the consumer surplus, and consequently the net value of the wetlands as waterfowl habitat.

It is quite important to recognize precisely which wetlands have been evaluated by capitalizing the annual consumer surplus associated with recreational hunting in Minnesota. The wetlands evaluated are all those responsible for generating the hunting experience in Minnesota. Clearly this includes the value of all of the recreational hunting areas in the state. The location of the breeding habitat involved is not quite so easily isolated, however. The value of all the breeding land in Minnesota is included in this capitalized value if, and only if, the waterfowl bred on Minnesota's marsh land are hunted exclusively in Minnesota. Clearly this is not the case. The birds are migrant, and some of the birds raised in Minnesota contribute to hunting further down the flyway. Consequently, a portion of the social value of the breeding component of Minnesota's marshes is not incorporated in the capitalized value of recreational hunting in Minnesota. Similarly, many of the birds hunted in Minnesota are bred in neighboring states and Canada. Thus, part of the capitalized value is assignable to breeding habitat outside the state.[10]

The interdependence of the value of hunting and breeding habitat[11] does not prevent separating the value of breeding land from the joint value of hunting and breeding land which we determined above.[12] The demand curve derived for hunter days has as a parameter a quality variable, Q, and a gravity variable, G. Both Q and G are partially dependent upon the number of waterfowl bred in and breeding in Minnesota. Presumably, if some breeding habitat were drained, fewer waterfowl would be attracted to the state in the spring; this would effect a reduction in both G and Q, and would shift the demand for hunter days to the left. Integrating under this new demand curve and capitalizing the value obtained provides an estimate of the value of the recreational hunting and breeding land involved, given the new,

---

[10] Surprisingly enough the practical difficulties of estimating that portion of the waterfowl kill in region i which was bred in region j is solvable with the help of an ingenious banding study conducted by Calvin Lensink. Detailed discussion of this problem is postponed until chapter 4. The problem of identifying exactly which breeding lands are included in the capitalized value of the recreational hunting arises because we have restricted ourselves to the Minnesota experience. This problem would not have arisen had we attempted to estimate the value of waterfowl habitat on a flyway or continental scale, for clearly all the breeding and recreational hunting land on the flyway would be included. Since preferences of hunters in different areas are not likely to be homogeneous, conducting the estimation on a flyway or continental scale may simplify the problem of identifying the relevant breeding land at the expense of differentiating between habitat values in different areas.

[11] The value of hunting land is influenced by the number of waterfowl attracted in the fall, and the value of breeding land is derived from the value which hunters place on hunting.

[12] Allen V. Kneese of Resources for the Future suggested the procedure for extracting the value of breeding land.

poorer quality of hunting in the state. The difference between the capitalized values before and after the drainage is the value of the lost birds (and consequently, breeding habitat) from the point of view of Minnesota hunters. This is an incomplete measure of the lost social value to the entire U.S. duck hunting population, because it does not include the losses due to poorer quality hunting further down the flyway.

The practical difficulties of estimating the change which occurs in the values of Q and G as a result of a certain amount of drainage are less formidable than they may seem at first. In the next chapter we develop and estimate an ecological model for waterfowl. The estimated equations may be used to predict the reduction in breeding population which occurs when wetlands of different types are reclaimed. One could predict the annual loss in ducklings brought to flight stage as a result of this reduction in the breeding population by a number of procedures, but the one that shows the most promise is the use of age ratios,[13] which would provide a crude estimate of the reproductive capacity of the breeding population. Finally, in order to estimate the influence of this reduction in the Minnesota resident bird population upon the quality of hunting, one has to have an estimate of the probability that a Minnesota resident bird will be shot in Minnesota during the year. Estimation of such a probability is feasible, and the problem is treated at length in the next chapter.

Thus, all the pieces necessary to estimate the value of waterfowl (on the wing and in the bag), recreational hunting land, and breeding habitat seem to be available, but our attempt to implement this model broke down in its earliest stages, with the estimation of the five relations whose general form is 3.1. Since these equations are used to derive the demand for hunter days, it is quite important that we develop a relation which fits the data well, or, in other words, which is able to explain much of the variation in the dependent variable, DH. Unfortunately, however, we are only able to explain 4 percent of the deviation in the dependent variable, i.e., $R^2 = 0.04$. Exactly why such

---

[13] The ratio of immature to adult waterfowl.

disappointing statistical results were obtained is extremely difficult to say. One possible cause is that for the year in which we had observations on C, Y, Q, and G, we had no information on DH, and had to simulate it based on the experience of the preceding year. I am afraid, however, that much of the responsibility for these results lies with the form of 3.1 itself. The relation as conceived is apparently too simple to explain the number of days hunted by individuals, and research will have to be done to identify variables that are more closely correlated to days hunted. Finally, the problems encountered in estimating 3.1 are purely statistical, and in no way reflect upon the suitability of the conceptual procedure developed for estimating wetland and wildlife values. The proposed technique is appropriate, but the model must be specified more fully.

Appendix to Chapter 3

Sampling Procedure and Description of the Data

The data with which I attempted to estimate the parameters in relation 3.1 were collected in three mail sample surveys of Minnesota waterfowl hunters. One survey was conducted by me, using a sample designed and drawn by the United States Department of the Interior,' Migratory Bird Populations Station at Laurel, Maryland (hereinafter referred to as the USDI), one by the USDI itself, and the third survey was conducted by the Minnesota Department of Conservation. The survey which I took requested information on the distance traveled by hunters from their home to their hunting area, their family income, and the amount of any rental payment or entry fee that they made for hunting rights.[14] A copy of the questionnaire appears at the end of this appendix.

The individuals in my sample had been surveyed the previous year by the USDI.[15] The USDI used a stratified random sampling procedure in which the sampling units were post offices, the strata consisting of four geographic zones in Minnesota and post office size categories. Ninety-seven Minnesota post offices were selected, and the postmasters were instructed to give a card supplied by USDI to each person who bought a federal duck stamp, and to ask the purchaser to mail the card with his name and address to the USDI. Hence, each individual in the sample supplied his name and address voluntarily. There were 5,349 hunters in the sample, or 4.82 percent of the estimated

---

[14] Early in this investigation it was thought that hunting land might be evaluated directly, using entry fees or their equivalent. A number of problems ultimately arose which led to abandoning this direction of inquiry. Foremost among these was the fact that less than 10 percent of the sample respondents indicated that they paid some kind of entry fee, and the results of a follow-up sample suggested that only a little more than 5 percent of the hunting population engages in such payments. Clearly, it would have been inappropriate to evaluate hunting land on the basis of the preferences of such a small percentage of the hunting population.

[15] One weakness in my sample and a possible cause of the disappointing statistical results is the fact that the sample was drawn from the 1963–64 hunting population, but information was requested about the 1964–65 hunting season. I justify this procedure on the grounds that a survey of these dimensions is extremely expensive, and given the limited resources available to me, no sample of comparable size and quality could have been drawn from the 1964–65 population.

111,000 potential adult hunters in Minnesota during the 1963–64 season.[16]

Among other variables the USDI collected information on the number of days that an individual participated in waterfowl hunting. Because the original formulation of my model did not call for such information, no question regarding frequency of participation appeared on the questionnaire which I sent to duck hunters. Subsequent revision in the model culminated in a form that employed days hunted as a crucial variable, and the number of days reported by an individual in 1963–64 was used as a proxy variable for participation in 1964–65. Past experience by the USDI indicates that there is an extremely high correlation between the number of days on which an individual hunted in one year and in the next.

Because it was felt that estimation errors would be introduced if respondents were requested to indicate the distance that they traveled to their hunting preserve, they were asked instead to list their home town and the town nearest their hunting area. Calculating the distance between every two towns in the state proved an excessively ambitious task. As a substitute for this, each hunter was assigned the air distance between the population center of the county in which he lived and that of the county in which he hunted most frequently. If a respondent hunted within his home county, he was assigned a nonzero number judged to be the average distance traveled within that county. These data were developed by John Borchert.

As a measure of the quality of the hunting in a county I used the mean number of ducks bagged per hunter day in that county during the 1964–65 season. Data to derive this statistic are collected annually by the Minnesota Department of Conservation (MDC). Attached to the Minnesota small game hunting license is a report card that requests, among other things, the county which the hunter patronized most frequently, the num-

---

[16] It should be noted that the sample is not drawn from the population of waterfowl hunters in Minnesota, but from the population of adult waterfowl hunters. Hunters under 16 years of age are not required to purchase a duck stamp, and consequently, no such hunters appear in the population sampled.

ber of ducks which he shot during the season, and the number of days of hunting. Submission of the card is mandatory, but the law is not enforced. Only 9,786 of the estimated 120,000 active hunters in 1964 returned the questionnaire. The MDC suspects that there is some upward bias in the data on the mean number of birds killed, because of evidence that hunters who have a relatively fruitless season neglect to report their experience. No attempt was made to correct this probable bias, its extent being unknown.

Summary of Response Statistics

The questionnaire was pretested by administering it to seven Twin Cities hunters and was then mailed to the remaining 5,342 hunters in the USDI sample. The response statistics are summarized below.

Sample size	5,342
Nonrespondent	2,331
Questionnaire returned undelivered	81
	2,412
Respondents	2,930
Void questionnaires	697
Correctly completed questionnaires	2,233
Rental payment made	289
No payment made for access	1,944

Questionnaire

1. In 1964, did you hunt ducks on PUBLICLY OWNED or PRIVATELY OWNED LAND or BOTH?

    public _____   private _____   both _____

Some duck hunters hunt on land which THEY OWN. Some hunters LEASE, directly from the owner of a slough or marsh, the RIGHT TO HUNT ON HIS LAND. Some hunters do not actually lease the land on which they hunt, but rather make a PAYMENT to the owner of ADJACENT LAND (an ACCESS FEE), in order to CROSS this land and get onto the hunting area. Finally, some hunters hunt without making any such payments. With these distinctions in mind, please answer questions 2 through 5.

2. In 1964, did you LEASE the right to hunt ducks on someone's land?

        yes _____   no _____

IF YES:
  a. HOW MUCH did you pay for the right to hunt on this land? (If you lease with a group of hunters, give only your SHARE of the lease payment.)

        $_____ per duck hunting season

  b. FOR THIS PAYMENT did you receive anything BESIDES hunting privileges?

        yes _____ (please specify)

        no _____

  c. General LOCATION of the LAND which you LEASED in *1964*:

      nearest town _____

      state _____

Did you hunt in the SAME place in *1963*?

        yes _____   no _____

IF NO, where did you hunt MOST FREQUENTLY in *1963*?

      nearest town _____

      state _____

  d. If you know the NAME and ADDRESS of the OWNER of the hunting land which you leased in *1964*, please PRINT it below.

name _____

street _____

town _____ state _____

3. In 1964, did you make an ACCESS PAYMENT allowing you to CROSS someone's land and get onto the area which you used for duck hunting?

        yes _____   no _____

IF YES:
  a. HOW MUCH did you pay for ACCESS privileges?

        $_____ per duck hunting season

  b. General LOCATION of this land:

      nearest town _____

      state _____

APPENDIX TO CHAPTER 3　　　　　　　　　　　　　　　　　　　　61

    Did you hunt in the SAME place in *1963*?

              yes _____　　no _____
    IF NO, where did you hunt MOST FREQUENTLY in *1963*?

        nearest town _____

           state _____

4. In 1964, did you hunt ducks on YOUR OWN LAND?

              yes _____　　no _____
    IF YES:
        General LOCATION of your land:

           nearest town _____

           state _____

5. In 1964, did you hunt on land which was NOT your own WITHOUT MAKING either ACCESS or LEASE PAYMENTS?

              yes _____　　no _____
    IF YES:
        a. General LOCATION of the land on which you hunted MOST FREQUENTLY:

           nearest town _____

           state _____

        b. Did you hunt on this land in *1963*?

                 yes _____　　no _____
        IF NO, where did you hunt MOST FREQUENTLY in *1963*?

           nearest town _____

           state _____

6. In order to know how far duck hunters travel to hunt, we have to know your CORRECT ADDRESS as of OCTOBER, *1964*:

           city or town _____

           state _____

7. Please indicate your approximate family income (before taxes) for 1964:

_____	0–1,499	_____	8,000– 9,999
_____	1,500–2,999	_____	10,000–14,999
_____	3,000–4,499	_____	15,000–24,999
_____	4,500–5,999	_____	25,000 or more
_____	6,000–7,999		

4

WETLANDS AS BREEDING HABITAT

In order to determine the optimal allocation of wetlands at the margin, the initial formulation of the problem called for comparing the discounted value of the income stream from an acre of arable land less the cost of reclamation ($V_{as} - I_s$) with the discounted value of the net benefits from an acre of wetlands in its natural state (V_{ws}). It was assumed that wetlands in their natural state have benefits only as waterfowl breeding and resting habitat and as recreational hunting regions. This chapter is devoted to the analysis of different types of marshes in order to determine their importance in the ecology of the waterfowl.

It will be recalled that it is a more difficult economic problem to determine the optimal allocation of wetlands as waterfowl habitat than as hunting environment, for it is conceivable that the distribution of hunting land (or the rights thereto) can be competitively organized, and an economically efficient situation established automatically by the market. But the migrant waterfowl that rest, breed, and are bred on wetlands not suitable for hunting purposes represent a noncapturable, nonmarketable resource to the marsh owner.[1] Hence there is a divergence be-

---

[1] It should be recognized that even if an owner of durable wetlands successfully realizes the recreational hunting value of his land, he may not be able to exploit his resource fully. Wetlands suitable for hunting purposes may also be important as breeding environment, and this component of the value of his property he cannot sell. Hence, even in the case of durable marshes, undervaluation is the most likely solution of the private market's operation.

tween private and social values. Such wetlands comprise an essential part of the waterfowl environment, but no part of this social value can be appropriated by the marsh owner.

The extent of the maldistribution of benefits and costs deserves emphasis. As we have seen, any social costs associated with maintaining these wetlands (planting in irregular patterns, crop depredation by migrating waterfowl, etc.) are likely to be borne by the marsh owners or their neighbors, and not by the duck hunting community. Hence, duck hunters derive all the benefits and sustain none of the production costs, while the wetlands owners (primarily farmers) receive none of the benefits and are saddled with all of the costs.

Although I was not able to establish a value for the output of breeding land (and hence, to evaluate breeding land itself), it is still important to consider the opportunity cost (in terms of lost waterfowl) associated with draining particular types of wetlands. Investigation of the agriculture sector indicated that, under competitive conditions, temporary wetlands (type 1 and possibly type 3) are the marshes most likely to be drained; the expense of draining more permanent marshes virtually precludes their reclamation. If it can be demonstrated that temporary wetlands play an insignificant role in the ecology of waterfowl, then the fact that these lands are in jeopardy is of no consequence. Given competitive prices for farm produce, no divergence exists between private and social costs and benefits in the agriculture sector, and no alternative output is forgone in the wildlife sector. Under these circumstances and in the absence of subsidies, drainage is socially desirable if it is privately profitable.[2]

If, on the other hand, temporary wetlands appear to be an important element in the birds' environment, then no such sweeping assertion emerges from the analysis. If marsh owners assign a zero price to the output of their wetlands, it is clear that at equilibrium an excessive amount of drainage will have been conducted. But, in the absence of an estimated value for

---

[2] See chapter 2, p. 22 and footnote 23, p. 34.

waterfowl, it is impossible to determine how severe the current misallocation of resources (if any) is between the two sectors.[3]

Methodological Design and Data Sources

It is important to know how sensitive waterfowl populations are to changes in their environment, especially to fluctuations in the inventory of temporary wetlands. It would be ideal if we had a closed ecological system that was representative of the breeding ground habitat located throughout Minnesota. Inferences regarding the entire waterfowl population could then be made on the basis of observations on this microcosm. The difficulty in designing such an experiment stems from the requirement that a *closed* ecological system must be maintained. Given the migratory character of waterfowl, it is virtually impossible to simulate an environment which produces the kind of information desired. The most that one can hope to demonstrate with data from an intensive study area is that waterfowl do not respond significantly to changes in certain elements in their environment. The significance of this type of negative result should not be discounted; as explained earlier, information of this character about type 1 wetlands would simplify our task enormously.

However, with data from a microecological unit it will not be possible to identify those critical components of the ecological universe which, if altered or eliminated, result in irreparable changes in the entire waterfowl population. For instance, if a certain environmental component were eliminated from the study area, and this resulted in markedly fewer birds nesting on the area in the future, it cannot be inferred that an identical change in the entire universe of breeding grounds would necessarily reduce the waterfowl population all along the flyway. The change which was effected in the study area made this region appear relatively less attractive than nesting areas available elsewere on the flyway. Hence, this change may merely have

[3] Society might want to reverse the current direction of the flow of resources by flooding land and transferring it to the wildlife sector, but without a price for waterfowl, there is insufficient information to support policy prescriptions of this character on economic grounds.

caused a redistribution of nesting waterfowl throughout the breeding grounds, and it cannot be concluded that either the waterfowl population or its reproductive capacity has been diminished. In order to gain any insight into the response of waterfowl populations to changes in their environment it will be necessary to examine aggregate data over a period of years, and consider waterfowl behavior in a general equilibrium setting.

The data employed in the analysis are all from secondary sources. The data for the microanalysis were developed by the United States Department of the Interior in the course of a lengthy, intensive study of the Waubay National Wildlife Refuge in South Dakota.[4] Waubay includes twelve sections of privately owned land (twelve square miles) and has an environment similar to that found throughout the prairie pothole regions of Minnesota, the Dakotas, and Canada.[5] Although Waubay's moderately rolling grassland is similar to that throughout the prairie pothole breeding grounds, it would be inappropriate to infer that the experience observed at Waubay would be duplicated exactly elsewhere.[6] Nonetheless careful consideration of these data is justified by Waubay's location in the principal nesting area and its similarity to the rest of the breeding grounds.

The aggregate analysis is based on data collected in the annual breeding ground surveys conducted by the Bureau of Sport Fisheries and Wildlife and an analysis thereof by Walter F. Crissey of that bureau.[7] Hunting and breeding activity

[4] I am indebted to Harvey K. Nelson, Jerome H. Stoudt, and John Lokemoen, all of the U.S. Department of the Interior, Fish and Wildlife Service, Northern Prairie Wildlife Research Center, Jamestown, North Dakota, for supplying me with these data. The data were collected over a 16-year period from 1952 to 1965.

[5] See Thomas A. Schrader, "Waterfowl and the Potholes of the North Central States," *Water: The Yearbook of Agriculture, 1955*, U.S. Department of Agriculture, Washington, D.C., pp. 598–602.

[6] Kahler Martinson, U.S. Department of the Interior, Migratory Bird Populations Station, Laurel, Maryland, in a private conversation.

[7] Walter F. Crissey, "Prairie Potholes from a Continental Viewpoint," *Saskatoon Wetlands Seminar*, Canadian Wildlife Service Report Series No. 6, Department of Indian Affairs and Northern Development, Ottawa, 1969, pp. 161–71.

together with a habitat inventory are recorded annually by the bureau for all the flyways throughout the United States and Canada.[8] The data are admittedly much less precise than those from the Waubay study, but they provide critical information nonetheless.

It will no longer be feasible to continue to distinguish wetlands simply as temporary or permanent marshes; slightly more technical classifications will be necessary. We shall adopt the Department of the Interior's definitions, given in footnote 3, chapter 2. Loosely one may consider that type 1 wetlands are "temporary," and types 3, 4, 5, etc., are "permanent," successively higher numbers indicating deeper and more durable marshes.[9]

The Contended Importance of Type 1 Wetlands

No dispute arises over the crucial importance of stable, durable, and relatively deep water areas as a factor in the environment of waterfowl. There is, however, considerable conflict regarding the importance of the temporary, satellite marshes (type 1 and occasionally type 3). Since these ponds retain water for such a short period, and are unlikely to be wet after midsummer one might think that they are completely unsuitable as breeding and nesting habitat. It is not the contention of conservationists that these ephemeral areas are important as nesting habitat, although some nesting does occur on them. Rather, a much more subtle and unique role is claimed for type 1 wetlands.

> ...a combination of many small and temporary water areas near other areas that hold water throughout the summer appears to be best for maximum waterfowl production. Even though these small water areas often dry out after a few days or weeks in the spring, they are essential to the maintenance of the waterfowl breeding population. This is because during

---

[8] *Waterfowl Status Report* (annual, 1958–65), U.S. Department of the Interior, Fish and Wildlife Service, Bureau of Sport Fisheries and Wildlife, Special Scientific Reports 40, 45, 51, 61, 68, 75, 86, and 90, Washington, D.C.

[9] There is no type 2 classification.

the breeding season mated pairs do not get along well with others of their species, and the many small wet spots provide the pairs with the privacy they apparently need to complete their courtship successfully.[10]

The hostility that drake waterfowl exhibit toward one another during the mating and breeding season is manifested in a practice called territorialization.[11] A drake chooses a breeding area and defends it against encroachment by other birds. Such aggressive tendencies are a widespread phenomenon in nature; the noted physiologist Konrad Lorenz has documented intraspecific competition for territory among many animals.[12]

Thus, a hypothesis advanced by a number of biologists and conservationists is that the drakes' territorial penchant during the early spring makes peripheral water areas a rather significant element in the birds' ecology. If wetlands are removed from the environment, male territorial behavior may prevent the displaced birds from crowding onto the remaining wetlands in order to breed. We must therefore consider the hypothesis that an abundance of type 1 wetlands in proximity to more permanent marshes[13] affects the size of the breeding population attracted to the area, for if it does, and if alternative regions are at their carrying capacity, then reclamation of type 1 wetlands could have a severe, deleterious impact upon waterfowl populations.

It is important to emphasize that the wetlands in question here are type 1 ponds in the neighborhood of more permanent

---

[10] Grady E. Mann, "Ducks and Their Water Home," *The Conservation Volunteer*, May–June 1957, p. 28.

[11] H. Albert Hochbaum, *The Canvasback on a Prairie Marsh* (Washington, D.C.: The American Wildlife Institute, 1944), p. 56. See also Schrader, in *Water*, p. 600.

[12] Konrad Lorenz, *On Aggression* (New York: Bantam Books, 1967).

[13] Wildlife experts readily concede that isolated, temporary wetlands have no value as waterfowl environment. "An isolated temporary pothole not near a more permanent pothole has no value because it furnishes no brood habitat and can satisfy only the courtship needs of the birds. On the other hand a permanent pothole by itself, while ideal for broods, is of limited value because only a few courting pairs will use it. Only when it is surrounded by a number of temporary waters will its maximum carrying capacity for broods be realized." Schrader, in *Water*, p. 601.

marshes. Isolated type 1 potholes have no opportunity cost in the wildlife sector. They dry up too soon to be useful as brood habitat, and their isolated location makes them of little importance as courting areas in the early spring. It follows from this that there is no divergence between private and social costs and benefits with respect to these lands, and consequently the competitive allocation of them is the social optimum. In addition, the only resources misallocated as a result of subsidy-induced reclamation of these lands are the resources employed in drainage and the marginal factors used for cultivation; the wildlife sector is independent of this activity.

Analysis of the Waubay Ecological Data

Although the question of central importance is whether the number of breeding pairs attracted to an area is influenced by the availability of type 1 wetlands, we are also concerned with the sensitivity of waterfowl breeding populations to fluctuations in other elements of their environment. For this reason regression analysis was applied to the comprehensive data developed by the USDI during its investigation of waterfowl environment and behavior at the Waubay National Wildlife Refuge. Data are recorded annually for each of the twelve sections on the number of breeding pairs and on the number and acreage of ponds of types 1, 3, 4, and 5. Although research at Waubay was initiated in 1950, data from some of the early years may not be comparable with more recent data because of changes in personnel and research procedures since the study's inception. Therefore, our analysis is confined to the years 1955 and 1957–65. (No data were collected in 1956.)

A model that attempted to explain the number of breeding pairs nesting in an area solely as a function of the habitat characteristics of that region would surely be ill-conceived. On an a priori basis at least, it would seem that the size of the aggregate waterfowl population migrating through a region in the spring is of critical importance in determining the number of birds which ultimately settle in the region during the breeding season. Failure to include this variable renders the model vulnerable to the criticism that it is an open system, subject to unexplained emigration and immigration of breeding pairs.

Although there is no measure of the spring migration through Waubay, an estimate of the waterfowl breeding population located in Minnesota and the Dakotas (henceforth referred to as the Tri-state breeding population) is available and serves as a useful proxy for the size of the spring migration. As mentioned earlier, aerial and ground crews employed by the Bureau of Sport Fisheries and Wildlife conduct extensive, annual surveys of the breeding ground population and ecology. Information is collected by region along each of the four flyways throughout the United States and Canada on the number of ponds in May and July, the waterfowl breeding population size, and the ducklings bred. The major purpose of these annual surveys is to estimate the relative size of the fall flight from each of the breeding areas, and hence the data are presented as indexes, and are not reliable as estimates of the absolute sizes of the variables.[14] Nonetheless, the data are comparable between years, and an index would seem to be sufficient for our purposes.

Before embarking on a statistical discussion of the models and their results, a word of caution seems appropriate. Although the results are encouraging, the models are a highly simplified means of describing the operation of a very complex system.[15]

[14] "The aerial crews count the birds on somewhat less than 1 percent of the total breeding area. This is sufficient coverage to reduce sampling error to less than 20 percent of the average population density in most survey areas, and to much less than 20 percent for the breeding range as a whole. The results of the breeding ground surveys are presented as *indexes*. When conducting aerial surveys of breeding birds or of broods, not all birds present are seen by the aerial crews.... Since there is no attempt to estimate the number of birds not seen, the indexes ... are based on birds actually seen, and it is emphasized that they do not constitute estimates of the total numbers present." *Waterfowl Status Report, 1963*, U. S. Department of the Interior, Fish and Wildlife Service, Special Scientific Report, Wildlife No. 75, Washington, D.C., October 1963, p. 5.

[15] The mind boggles at the dimensions of the task involved in developing a comprehensive model for predicting the number of breeding waterfowl attracted to an area. Types and densities of vegetation, bacterial content of the water, the nature of the surrounding topography, and proximity to farmland might all be important variables. Not only is data not available for these variables, but one might be hard pressed even to propose measures for some of them.

Statistical models are designed in large part on a pragmatic basis; the variables included are those which the researcher considers to be influential and for which data are available. The models developed and tested in this study are no exception. They are attempts to identify the relative importance of different environmental factors in attracting breeding waterfowl, but no claim is made that they are exhaustive as descriptions of the relevant environment. Indeed their comprehensiveness and precision are revealed by the size of the R-squares and the statistical significance of the coefficients estimated.

Since there appeared to be no a priori grounds for choice between the significance of numbers of wetlands and acres of wetlands as explanatory variables of the size of the breeding population, two models were proposed and tested:

$$B = f(W_1, W_3, W_{45}, P) \text{ and}$$
$$B = f(A_1, A_3, A_{45}, P),$$

where
- B = breeding pairs of waterfowl,
- W_i = wetlands of type i (measured in number of ponds),
- W_{ij} = wetlands of type i through j inclusive,
- A_i = acres of type i wetlands,
- A_{ij} = acres of wetlands of type i through j inclusive,
- P = index of the Tri-state waterfowl population (in 1,000's of birds),
- b_i = the regression coefficient of the variable with subscript i, and
- b_p = the regression coefficient of P.

Despite the fact that the data are time series data, the initial approach to the problem involved grouping the data as if they were 120 observations from one cross-sectional sample, i.e., one observation on each of twelve sections for ten years.

Statistical inference has meaning only if the sample data are drawn from the same or identical populations. Otherwise, one is either making inferences about a population that does not exist as an identifiable entity, or one is combining populations with no knowledge of their relative weights. In order to avoid spurious conclusions, then, it is important that the twelve sections at Waubay display statistical properties which indicate that they come from the same population, and may be grouped for analytic purposes.

It was suggested that, before proceeding with the analysis, sectional differences be tested for by fitting regressions of the above models to the data for each individual section, comparing the regression residuals to be certain that they were similar.[16] If two sections have residual sums of squares (SS_R) whose magnitudes differ markedly, this may be taken as prima facie evidence that the populations from which these two sections were drawn are statistically dissimilar in one of their basic parameters (their variance). Since homogeneity of variance is a necessary prerequisite for combining data, the sections could not be grouped.

The results of these tests indicate that four of the sections at Waubay have statistical properties very different from the other eight and from each other, and consequently cannot be incorporated into the analysis at all.[17] A more complete exposition of these statistical results together with an attempt to identify those environmental features which differentiate these four sections from the rest of the preserve appears in the appendix to this chapter.

Regressing breeding pairs against acres of wetlands and the Tri-state waterfowl population index for the remaining eight sections produces the following equation:

$$B = -7.3754 + 1.8567\ A_1 + 1.1721\ A_3 \qquad (4.1)$$
$$(0.5652) \qquad (0.2883)$$
$$+\ 0.5230\ A_{45} + .0124\ P$$
$$(0.0549) \qquad (0.0088)$$

$N = 80$, $R^2 = 0.781$, $F(4, 75) = 67.1778$ (significant at 99.95%). From a statistical point of view the result is encouraging because of the goodness of fit and the high level of significance associated with the coefficients of the variables. We are able to account for 78.1 percent of the deviation in the dependent

---

[16] I am indebted to Donald Bentley of the Pomona College Mathematics Department for this suggestion and his critical comments on the ensuing analysis.

[17] Remarkably enough, even if one does include all of the sections in the analysis, the statistical results remain virtually unchanged. For purposes of

variable, and all of the coefficients in equation 4.1 have the appropriate sign, with b_1, b_3, and b_{45} being significant at the 99 percent level. The coefficient of P is significant only at the 80 percent level, but even if b_p were statistically significant at an acceptable level, its absolute value would make P unimportant from a practical point of view.[18]

The small size of b_p does not necessarily mean that the aggregate waterfowl population is unimportant as a determinant of the number of birds breeding at Waubay. It may or it may not be. The proper interpretation is that, given the current levels of Tri-state waterfowl populations (ranging from a low of approximately 1 million in 1959 to approximately 3 million in 1963), fluctuations in the wetland acreage at Waubay have a greater influence on the number of birds that nest there than do fluctuations in P.

One possible explanation for the small size of b_p and the apparent lack of importance of the aggregate waterfowl population in determining the size of the breeding population at Waubay is that Waubay is located in the southern portion of the continental breeding grounds, and migrating waterfowl reach Waubay early in their flight north. It may be that many

---

comparison with equation 4.1 the regression of breeding pairs against acres of wetlands and the Tri-state waterfowl population index for all twelve sections is given here:

$B = -12.0154 + 1.9975\ A_1 + 1.2273\ A_3 + 0.5313\ A_{45} + 0.0167\ P$
$\quad\quad\quad\quad\quad (0.6326)\quad\quad (0.1250)\quad\quad (0.0605)\quad\quad (0.0090)$

$N = 120$, $R^2 = 0.739$, $F(4, 115) = 81.6833$ (significant at 99.95%).

[18] It may be that multicollinearity between P and the other variables is preventing us from estimating b_p with more precision. Relative to the simple correlations between the other independent variables P is rather highly correlated with the acreage variables. Although in absolute size these correlations are small, ranging from 0.316 to 0.577, jointly they may be sufficient to cause b_p to be insignificant.

Since P is measured in 1,000's of birds, the interpretation of this coefficient is that for every 100,000 additional birds in the Tri-state population index, 14.88 breeding pairs (1.24 pairs per section × 12 sections) nest at Waubay. Thus, on the average it takes an 11 percent fluctuation in the Tri-state population to produce a 3 percent fluctuation in the number of breeding pairs at Waubay.

birds select the southernmost habitat first for nesting purposes, and that latecomers are forced to fly farther north to find suitable habitat. If this description of waterfowl behavior in the selection of nesting habitat has any validity, then even very large fluctuations in the size of the spring migration might not influence the number of birds that settle in the south.

There is disagreement among waterfowl biologists on this point, but some evidence does exist to corroborate this portrayal of the dynamics of nesting in the prairie pothole region. Although homing, the tendency of female waterfowl to return to the same or nearby nest sites year after year, is a well-documented characteristic, birds will pioneer a new breeding place if the old one proves unsuitable, and, in some instances, if a new one appears more attractive. This contention is supported by the fact that waterfowl that nest at southern latitudes are very vulnerable to the gun, and subject to unusually high mortality rates. Nonetheless, the breeding grounds fill up year after year, an indication that pioneering must be occurring.[19]

All this is of secondary interest for the moment; the statistic to focus on here is b_1. It is highly significant, positive, and relatively large in magnitude. This constitutes rather strong empirical support for the contention that waterfowl are more highly attracted to areas with type 1 wetlands than to areas that do not have such satellite ponds. The estimate of the extent of this attracting quality is 1.8567 additional breeding pairs for every additional acre of type 1 wetlands (given the levels for all other variables). The application of this value for b_1 to all other areas throughout the breeding grounds is probably injudicious, but this is not the central point. The qualitative result that type 1 wetlands are an important attracting element in waterfowl habitat is the salient feature of the analysis, for it means that depletion of type 1 wetlands may have an opportunity cost in

---

[19] It is important to note that if the waterfowl population which would normally "home" to a region is depleted (say due to heavy exposure to hunting), the vacated potholes will fill to some extent with pioneers, but not to the level at which they would normally have been occupied. (Harvey K. Nelson and Walter F. Crissey in private communications.)

the wildlife sector. The particular value of the estimate is of subordinate consideration.

At the risk of being repetitious, it is important to emphasize again the restraint that must be observed in interpreting this result. It is not legitimate to project the outcome of this analysis of a microenvironment to the entire continental breeding grounds. Reducing the inventory of type 1 wetlands throughout the flyway may or may not diminish the waterfowl population. This problem can be illuminated only by the examination of aggregate data, a task to which the next section is devoted.

Although wetlands' acreage and number of ponds are highly correlated, exception could be taken to the preceding analysis on the grounds that ponds, not acres, are the relevant measure. Substituting number of ponds (W_i) for acres (A_i) in the model produces a relation which, although generally similar to equation 4.1, is clearly inferior to it for predictive purposes.

$$B = -18.5269 + 1.3626\ W_1 + 0.5116\ W_3 \qquad (4.2)$$
$$(0.3431)(0.5387)$$
$$ + 3.4619\ W_{45} + 0.0302\ P$$
$$(0.6523)\phantom{W_{45} +\ }(0.0103)$$

$N = 80$, $R^2 = 0.667$, $F(4, 75) = 37.5209$ (significant at 99.95%).

The coefficients in equation 4.2 all exhibit the expected sign, with b_1, b_{45}, and b_p being significant at the 99 percent level or better. But, with equation 4.2 only 66.7 percent of the deviation in the dependent variable can be explained, as opposed to 78.1 percent with equation 4.1. In addition to this, b_3 is very insignificant statistically, a result due probably to multicollinearity between W_3 and the other independent variables.[20] However, the size and statistical significance of b_1 in equation 4.2 do reinforce our earlier conclusion that waterfowl consider regions with type 1 satellite ponds more attractive than those without such peripheral water areas.

---

[20] W_3 has simple correlation coefficients ranging from 0.525 to 0.578 with the other arguments; none of the other variables has simple correlations as high as this. As in the case of b_p in equation 4.1, the absolute sizes are not alarming, but jointly they may be sufficient to prevent precise estimation by b_3.

Aggregate Analysis

The reason for examining waterfowl and their environment was to illuminate the degree to which changes in habitat influence the size of waterfowl populations. The investigation of the Waubay Refuge showed that the number of waterfowl attracted to an area is affected by both the amount of water available and the composition of the water areas; wetlands of varying durability seem to be an important characteristic of desirable habitat. Despite all this, it was impossible to resolve the more pressing problems: How do waterfowl behave when wetlands throughout the flyway are depleted? Are they willing to crowd onto the smaller habitat area, or is the territorial instinct so intense as to prevent greater concentration of the population? Does increased density alter reproductive capacity?

As mentioned previously, data as precise as that collected at Waubay are not available on an aggregate scale. In the course of its breeding ground survey the USDI does estimate the number of ponds, the size of the waterfowl population, and the number of ducklings bred, but no attempt is made to classify water areas by wetland type. Nonetheless a remarkable amount of understanding of the mechanics of migratory waterfowl behavior can be extracted from this data.

Walter F. Crissey has shown that there is a strong inverse correlation between the percentage of the continental waterfowl population present in the northern breeding areas (northern Alberta and the western portion of the Northwest Territories) and the number of water areas in the southern portions of the Prairie Provinces (Manitoba, Saskatchewan, and Alberta) in the spring. Simultaneously, and not unexpectedly, there is a strong positive relationship between the percentage of the continental population located in the southern parts of the three Prairie Provinces and the number of water areas there.[21] The implica-

---

[21] Crissey, "Prairie Potholes from a Continental Viewpoint," p. 161. The water areas in the Dakotas, Minnesota, and Montana were not incorporated into this analysis, because a comparable water count did not exist for these regions for the period covered by the analysis. It is known that the trend in water areas in these regions complemented that in the Prairie Provinces, and consequently the results of the analysis are almost certainly

tion of these findings is that if the habitat in the south is depleted (by drought or drainage), migrating waterfowl will continue north in search of satisfactory nest sites rather than crowd onto the southern breeding grounds. If it had been found that the breeding population in the southern Prairie Provinces was largely independent of the number of wetlands there, one could doubt the validity of the territorial hypothesis as a description of waterfowl behavior. The evidence presented by Crissey, however, supports the claim that birds will not crowd, and that drainage of habitat in the south drives the birds north to breed. (The Canadian provinces and the prairie pothole region are shown on the map facing page 1.)

Since the aggregate data are expressed in terms of water areas and not classified by wetland type, there is no direct way of estimating the effect that fluctuations in type 1 wetlands alone have upon the breeding population located in the Prairie Provinces. However, since the type 1 and possibly type 3 wetlands are the ones most vulnerable to both drought and drainage, it would seem a logical corollary of Crissey's observations that reductions in the number of temporary wetlands in the prairies forces the waterfowl north in search of nesting habitat. Hence, the aggregate data seem to corroborate the conclusions arrived at from consideration of the Waubay data: Type 1 wetlands are an important element of nesting habitat, and in their absence the birds move elsewhere.

At this point it might be argued that waterfowl behavior has still not been examined in a general equilibrium setting. The prairie breeding grounds are only a portion of the continental habitat, and if there is sufficient northern habitat to sustain the waterfowl population, then it can be maintained that depletion of the potholes in the prairies has no opportunity cost in the wildlife sector.

---

applicable to the breeding grounds in the north-central United States as well. The regressions demonstrating the relationship between the water areas in the southern parts of the Prairie Provinces and the percentage of the continental waterfowl population located in the northern and southern Canadian breeding grounds are in the appendix to this chapter.

For two reasons the proposition that the northern breeding grounds may serve as an ideal substitute for the prairie habitat does not withstand scrutiny. First, the quality of hunting in Minnesota and the Dakotas would be significantly impaired by such a substitution of breeding locale. (This topic is developed in the next section of this chapter where it is shown that local breeding grounds supply a large percentage of the duck harvest in most instances, and that birds bred or nesting in an area remote from the point of harvesting have a smaller probability of being shot.) Secondly, Crissey is able to demonstrate that waterfowl have been less prolific in years when the breeding population has been forced to go north in order to find suitable habitat.[22]

The northern habitat seems perfectly capable of supporting adult birds during the summer, but it does not appear to be a good substitute for the prairie potholes as a breeding ground. There is not sufficient data available to examine all species, but if the number of mallard young produced in North America (Y) is regressed against the number of July ponds in the southern parts of the Prairie Provinces (W) for the years 1955–65, 50.8 percent of the variation in number of young produced can be explained.

$$Y = 5.4274 + 2.2155\ W \qquad (4.3)$$
$$(0.7272)$$
$N = 11$, $R^2 = 0.508$, $F(1, 9) = 9.2822$ (significant at 95%).

More important, the coefficient of W is statistically significant at the 99 percent level, and exceedingly large in size. Since the data used here are estimates of the absolute values of the variables, not just indices, the interpretation of this coefficient is that for every million additional July ponds in the Prairie Province breeding grounds 2.2155 million mallard young are produced. If what appears to be an extreme observation (1957) is eliminated from the regression, an even more significant

---

[22] The data for the following analysis (equations 4.3, 4.4, and 4.5) come from Crissey, "Prairie Potholes from a Continental Viewpoint," figures 1 and 4, pp. 162 and 164.

result is obtained, with 83.7 percent of the dependent variable being explained.

$$Y = 4.8744 + 2.1049\ W \quad\quad (4.4)$$
$$(0.3277)$$

$N = 10$, $R^2 = 0.837$, $F(1, 8) = 41.2552$ (significant at 99%).

Corroboratory evidence for the crucial importance of the prairie potholes in preserving the size of the North American waterfowl population[23] is obtained by regressing the entire continental duck breeding population (P) in year $t + 1$ against the number of July ponds in the southern parts of the Prairie Provinces (W) in year t for the period 1954–65. According to this regression, for every million additional July ponds in the Prairie Provinces there are 4.5139 million additional ducks in the continental breeding population the following spring. This relation accounts for a remarkable 79 percent of the variation in P, and the coefficient of W is significant at the 99.9 percent level.

$$P = 37.8194 + 4.5139\ W \quad\quad (4.5)$$
$$(0.7360)$$

$N = 12$, $R^2 = 0.790$, $F(1, 10) = 37.6162$ (significant at 99%).

The correlation between the number of water areas in the prairies and the size of the waterfowl population may, of course, be entirely spurious. If the inventory of water areas in the northern breeding grounds experienced fluctuations that coincided with those in the south, then the above analysis would indeed be deceptive because W, the prairie potholes, would then have been serving merely as a proxy variable for the water areas throughout the continent. To attribute special significance to the southern breeding grounds under these circumstances would have been fallacious. However, since drought is a virtually unheard of phenomenon in the northern breeding grounds, our results remain unweakened.[24]

---

[23] Excluding scooters, eiders, mergansers, old-squaw, black ducks, and wood ducks.

[24] Although the evidence indicates that the waterfowl population reproduces at a lower rate when it is forced north, the possibility remains that the birds might ultimately acclimate themselves to the northern habitat and reproduce at the same rate in the north as they do in the south. The

No one is sure what accounts for the inferiority of the northern breeding grounds as reproductive habitat. Crowding appears to be an unlikely possibility; the number of water areas in the north is estimated in the trillions, which makes the ratio of ponds per breeder far greater than that elsewhere. This does not completely exclude crowding as a possible explanation, however, for many of the northern water areas may be unsuitable as nesting habitat. No one knows for sure. One element that undoubtedly contributes to the lower productivity of the waterfowl population in the north is the shorter length of the breeding season. Farther south in Canada and in the United States the weather is such that renesting can occur if the initial attempt to raise a brood fails. In the north the season is too short to permit this.[25] Whatever the cause, the evidence indicates that the northern breeding grounds have a considerably lower capacity than those in the south for maintaining the continental waterfowl population.

Implications of Breeding Habitat Depletion Upon Hunting Harvest

Thus far this chapter has assessed the effect of changes in the composition and location of waterfowl habitat on the size of breeding populations and their productivity. It has been demonstrated (1) that type 1 and type 3 wetlands—the wetlands most vulnerable to drought and drainage—are an important component of the birds' environment, and that in their absence the birds move elsewhere; (2) that when the number of water areas in the southern breeding grounds is lowered, the birds migrate farther north to nest rather than crowd onto the remaining southern habitat; and (3) that the continental waterfowl population is significantly less prolific when forced north. Having strayed somewhat from a regional to a continental perspective, we now return to the case of particular concern, Minnesota.

---

possibility seems remote, however, and permitting the continued depletion of the southern potholes until the question is resolved might prove a very expensive experiment.

[25] Walter F. Crissey, in a private conversation.

In light of the evidence that wetlands, both durable and temporary, in the southern Prairie Provinces and north-central United States enjoy a position of unique importance as waterfowl habitat, the conclusion of the previous chapter that many of the U.S. water areas are in jeopardy as long as price supports for farm commodities are retained takes on added significance.[26] Nonetheless, this does not constitute sufficient evidence to conclude that the quality of duck hunting in Minnesota will suffer severely unless early administrative action is taken to forestall drainage of waterfowl habitat within Minnesota. Abstract from the value of wetlands as hunting facilities, and concentrate exclusively on their characteristics as breeding grounds. Despite the demonstrated relevance of types 1, 3, 4, and 5 marshes as breeding habitat, one might still challenge the contribution of Minnesota's wetlands to the state's fall hunting harvest on the grounds that "about 80 percent of the ducks [on the continent] are raised north of the Canadian boundary."[27]

Hence, in order to assess the possible consequences of drainage of Minnesota's wetlands on the quality of hunting, it is crucial to know what percentage of the waterfowl bagged (killed and retrieved) in Minnesota is composed of resident birds.[28] Let event S equal a Minnesota resident bird bagged this year,

---

[26] It will be recalled that all temporary wetlands in our sample are vulnerable as long as price supports are in effect. Also, unless hunters are willing to pay a considerable rental fee for the use of permanent wetlands in the fall, current price supports are sufficient to make drainage of these lands profitable in the fertile southern regions of Minnesota. In addition, our contention that drainage of permanent potholes in northwestern Minnesota does not appear warranted under present conditions was tempered by the qualification that if a ditch alone is sufficient for adequate drainage, then even these less fertile lands are likely candidates for reclamation. It should be emphasized further that because type 3 wetlands normally go dry in the summer, there is no possibility that they will generate any rental fees from hunting, and they therefore enjoy a less secure position than type 4 and 5 ponds.

[27] Forrest B. Lee et al., *Waterfowl in Minnesota*, Minnesota Department of Conservation, St. Paul, Minnesota, 1964, p. 55.

[28] A resident bird is defined as either a member of the breeding population in the state during the year in question or an immature, i.e., a duckling brought to flight stage. All other birds will be termed imports.

and event S_m equal a bird bagged in Minnesota this year. Then,

$$P(S \cap S_m) = P(S) \, P(S_m|S)$$
= the probability that a Minnesota resident bird will be shot within the state this year.

Letting W equal the population of waterfowl in the fall flight from Minnesota, i.e., adults plus immatures, we obtain

$$K_m = P(S \cap S_m) \, W$$
= the number of resident birds killed in Minnesota.

Estimation of the unknowns in this last relation is complicated by a number of factors. (1) The Minnesota Department of Conservation estimates of the waterfowl population in Minnesota are the best available but are still not refined. (2) Immatures are much more vulnerable to hunters than adult birds and this makes it necessary to distinguish between age groups when specifying W and $P(S)$. (3) Banding studies indicate that certain species of birds display a higher probability of getting shot than others. Hence W and $P(S)$ must be classified by species as well.

A banding study conducted by the Minnesota Department of Conservation provides data on a_{ij}, the percentage of bird bands returned for species i, age j, following the first hunting season after banding occurred.[29] Due to a severe nonresponse bias on the part of hunters who retrieve banded birds, a_{ij} cannot be equated with the annual probability that a bird of that particular species and age will be shot. Studies of response bias indicate a remarkably uniform behavior by hunters, with approximately 40 percent of the bands being returned.[30] Therefore, correcting a_{ij} for nonresponse bias, we have

$$P(S_{ij}) = 2.5 \, a_{ij}$$
= the probability that a Minnesota resident bird of species i, age j will be shot this year.

---

[29] Lee et al., *Waterfowl in Minnesota*, p. 86.

[30] Aelred D. Geis and Earl L. Atwood, "Proportion of Recovered Waterfowl Bands Reported," *Journal of Wildlife Management*, April 1961, p. 158. Also, Frank C. Bellrose, "A Comparison of Recoveries from Reward and Standard Bands," *Journal of Wildlife Management*, January 1955, pp. 71–75.

A major banding project conducted by Calvin Lensink provides sufficient data to estimate $P(S_m|S)$.[31] Of 1,266 Minnesota bands returned, 787 or 62.2 percent, were recovered within Minnesota. We now have

$$K_m = \sum_i \sum_j P(S_{ij}) P(S_m|S) W_{ij}$$

$$= 0.622 \ (2.5) \sum_i \sum_j a_{ij} W_{ij}.$$

The Minnesota Department of Conservation's most recent estimate (1966) of the breeding pair waterfowl population by species appears in table 8. Substituting the appropriate values

TABLE 8. MINNESOTA WATERFOWL POPULATION ESTIMATES AND BAND RECOVERY DATA

| Species | Number of breeding pairs [a] | Percent of bird bands returned following the first hunting season after banding occurred [b] ||
		Adults	Immatures
Mallard	100,000	10.0	18.9
Blue-winged teal	150,000	2.4	5.5
Ring-necked	75,000	10.5	17.6
All others	100,000	10.0 [c]	13.3 [c]

[a] This information was provided by Robert L. Jessen of the Minnesota Department of Conservation in August 1966. An estimate of the waterfowl population is obtained by first multiplying the number of breeding pairs by 2 in order to calculate the number of adults. The number of immatures produced is estimated by multiplying the number of adults by 1.2. Adding adults to immatures gives the total population size.

[b] Forrest B. Lee et al., *Waterfowl in Minnesota*, Minnesota Department of Conservation, 1964, p. 86.

[c] Unweighted, arithmetic mean for "other" species.

into the above relation, we obtain $K_m = 298{,}109$. The total waterfowl bag in Minnesota in 1964 was 1,151,000 according to the Department of the Interior's survey of hunters.[32] Hence, it is estimated that the resident bird content of Minnesota's duck bag is around 26 percent.

[31] Calvin J. Lensink, *Distribution of Recoveries from Bandings of Ducklings*, Special Scientific Report, Wildlife No. 89, U.S. Department of the Interior, Washington, D.C., 1964.

[32] *Waterfowl Status Report, 1965*, p. 102.

It would be somewhat hazardous to infer from this that instant conversion of the state's breeding facilities to farmland would mean a 26 percent reduction in ducks shot within the state.[33] Although waterfowl display a strong homing instinct, the displaced Minnesota birds would adjust to the loss of their traditional nesting grounds by dispersing throughout the prairie pothole regions of the Canadian provinces and the adjacent states.

From the data collected in the Lensink banding study, it is possible to estimate $P(S_m|S_i)$, the probability that a bird migrating south from region i will be shot in Minnesota, given that the bird is going to be shot. Estimated values for $P(S_m|S_i)$ for the most relevant areas appear below.

	$P(S_m|S_i)$
Alberta	0.037
Manitoba	0.089
Saskatchewan	0.032
North Dakota	0.134
South Dakota	0.163

Even in the most optimistic case, where all the birds move to South Dakota, $P(S_m|S_i)$ is reduced from 0.622 to 0.163. Performing the relevant calculations yields the estimate that the annual waterfowl harvest in Minnesota would be reduced by 19.1 percent if the displaced birds settled in South Dakota, and by 22.2 percent if they dispersed evenly throughout the above five regions.

These are not insignificant magnitudes, and it would appear that the opportunity cost (in terms of waterfowl bagged) of forcing neighboring regions to produce game birds for Minnesota hunters by massive depletion of the nesting habitat in the

---

[33] Note that it is still assumed that only breeding facilities are removed. However, some of these wetlands also serve as hunting sloughs, and their depletion will mean either fewer birds attracted to the state in the fall or crowding of the fall migration onto large lakes and river basins (and, consequently, crowding of the hunters, too). These events will further reduce the number of birds killed, an impact that is not accounted for above.

state is not trivial.[34] These results are not surprising, and the Lensink banding study presents convincing evidence that they can be generalized to other regions. Loss of local habitat throughout the north-central states and the south-central Canadian provinces will force the waterfowl population to transfer to the northern breeding grounds, an event that will further impair hunting quality in the midwest because the reproductive rate is lower in the north.

Economic Implications

Without an estimated value for waterfowl it is impossible to derive unqualified conclusions and provide precise recommendations concerning the optimal distribution of all classes of wetlands between the agriculture and wildlife sectors. No difficulty arises regarding the allocation of permanent marshes. In the previous chapter it was found that the investment cost involved in reclaiming permanent wetlands precludes drainage, if perfectly competitive prices prevail in the agriculture sector. Even if society considered waterfowl to be valueless, drainage of these lands would be uneconomic. It was also concluded that isolated type 1 wetlands have no waterfowl value. The operation of competitive prices in the agriculture sector would determine an optimal allocation of these marshes. The area of unresolved conflict centers around type 1 wetlands near more permanent water areas. In chapter 2 it was found that even under competitive conditions it proved privately profitable to drain some of these satellite marshes. In this chapter it has been shown that these marshes are an important element in the birds' environment. Drainage therefore entails an opportunity cost. This opportunity cost remains unevaluated, because the value of a duck could not be estimated. Because there is no market mechanism by which type 1 marsh owners can appropriate the value of their resource when it is devoted to waterfowl production, they impute a zero price to the output of their wetlands,

---

[34] As the previous footnote suggests there are other, unaccounted for reductions in K_m which would ensue from such drainage. Therefore, the above estimates are probably conservative.

and at equilibrium an excessive amount of drainage will have been undertaken. We have no way of knowing how severe this misallocation of resources is.

Summary

1. Isolated type 1 potholes not near durable water areas or large complexes of type 1 marshes like those found in southern Minnesota have no opportunity cost in the wildlife sector. Consequently there is no divergence between private and social costs and benefits with respect to these lands; the competitive allocation of them is economically efficient. That these lands may be in jeopardy of being reclaimed is of no economic consequence (under competitive conditions in the agriculture sector), for no alternative output is forgone in the wildlife sector when these marshes are converted to arable land. The only resources misallocated as a result of subsidy-induced reclamation of these lands are the resources employed in drainage and the marginal factors used for cultivation; the wildlife sector is independent of these activities.

2. Breeding waterfowl are more highly attracted to habitat areas with type 1 wetlands adjacent to the more permanent ponds than to regions which do not have such satellite marshes. Consequently, drainage of such type 1 marshes is likely to induce waterfowl to seek nesting sites elsewhere.

3. If the habitat in the southern breeding grounds (southern Manitoba, Alberta, and Saskatchewan, and the north-central United States) is depleted, migrating waterfowl will continue north in search of satisfactory nesting sites rather than crowd onto the reduced southern habitat. This transfer of the breeding waterfowl to the northern breeding grounds is significant because the continental waterfowl population is markedly less prolific when forced north, and consequently the population size is not maintained.

4. In order to assess the possible consequences of drainage of Minnesota wetlands on the quality of hunting, it is crucial to know what percentage of the waterfowl bagged in Minnesota is composed of resident birds. With the help of banding studies it was possible to estimate that the resident bird content of

Minnesota's duck bag is around 26 percent, and that massive depletion of the nesting habitat in the state, forcing neighboring regions to produce game birds for Minnesota hunters, would reduce the fall harvest in Minnesota by approximately 22 percent.

This estimate of the reduction in the fall harvest in Minnesota is a lower bound, because it has been generated under the assumption that although drainage is occurring in Minnesota, the inventory of wetlands throughout the rest of the flyway is constant. Clearly, if wetlands' reclamation occurs concurrently in adjacent areas, the impact upon waterfowl populations and hunting quality both in Minnesota and elsewhere on the flyway will be a great deal more severe. Birds whose nesting habitat in Minnesota is removed will not be able to pioneer and resettle in adjacent areas with the same ease, since habitat in these areas has also been depleted.

5. An excessive amount of drainage of satellite type 1 marshes will occur under competitive conditions in the agriculture sector. The severity of this misallocation of resources cannot be assessed without an estimated value for waterfowl.

Appendix to Chapter 4

Environmental Differences at Waubay

Regressions of the principal models were fitted to the data for each of the sections at Waubay, in order to determine whether the residuals were sufficiently similar to justify grouping the data. The sum of squares of the residuals (SS_R) for eight of the sections was very similar, but the remaining four had SS_R which ranged from four to seven times larger than the average for the eight with similar residuals. The differences between the SS_R for these four sections and the SS_R for the other eight were large enough to raise serious doubts about the statistical similarity of the populations from which they were drawn. Within the group of four the same problem arose: the SS_R differed so widely as to preclude combining the data from these four sections, and running a statistical analysis on them jointly. Hence, it is not legitimate to consider the data as one cross-sectional sample of 120 observations; only the eight sections whose residuals are similar may be grouped and analyzed jointly.

Although formal regression analysis cannot be applied to four of the twelve sections, an attempt was made to identify the environmental features that might differentiate these four sections from the rest of the preserve, and lead to the disparity in residuals.[35] Some differences do emerge, but they are not universal enough to support unequivocal conclusions. All four sections have more of their acreage devoted to the cultivation of small grains and less to grazing than the average for the rest of the preserve.[36] It may be that more intense cultivation of small grains is a factor that influences the birds, and renders these sections less attractive as nesting habitat. One might feel more secure about this contention, if the distinction regarding land use were more pronounced. However, some of the other sections at Waubay are cultivated just as intensively. Although the four sections may be somewhat flatter than the rest of the area, the

---

[35] I wish to thank Harvey K. Nelson and the other researchers at the Northern Prairie Wildlife Research Center, Jamestown, N.D., for their meticulous and laborious efforts in developing the data for this analysis, and for indicating possible interpretations of the statistics.

[36] Although Waubay is a wildlife preserve, the land is all privately owned and almost entirely agricultural.

topographical differences are not severe. The four maverick sections have more type 3 and less type 1 wetlands than the average for the other eight, and the type 3's seem to be of the larger, shallower variety. Again, however, counterexamples can be found among the other eight sections, thereby making unconditional conclusions precarious.

Statistical Relationships between the Number of Prairie Ponds and the Location of the Waterfowl Population

Let P_s and P_n be the percentage of the continental waterfowl population located in the southern portion of the Prairie Provinces and the northern breeding grounds (northern Alberta and the western portion of the Northwest Territories) respectively, and let W_m and W_j be the number of ponds (in millions) in the southern portion of the Prairie Provinces in May and July respectively. The data cover the eleven-year period 1955–65.

$$P_n = 35.9773 - 4.7038 \ W_m \quad (4.6)$$
$$(1.1980)$$

$N = 11$, $R^2 = 0.632$, $F(1, 9) = 15.4169$ (significant at 99%).

$$P_n = 30.0833 - 4.4916 \ W_j \quad (4.7)$$
$$(1.2176)$$

$N = 11$, $R^2 = 0.602$, $F(1, 9) = 13.6078$ (significant at 99%).

$$P_s = 18.5723 + 7.5348 \ W_m \quad (4.8)$$
$$(2.2314)$$

$N = 11$, $R^2 = 0.559$, $F(1, 9) = 11.4022$ (significant at 99%).

$$P_s = 26.5833 + 7.9167 \ W_j \quad (4.9)$$
$$(1.9577)$$

$N = 11$, $R^2 = 0.645$, $F(1, 9) = 16.3526$ (significant at 99%).

We can see from the regressions that a reduction of one million ponds in the southern Prairie Provinces results in approximately a 4.5 percent increase in P_n and a 7.5 percent decrease in P_s. The coefficients are all significant at the 99 percent level. Again, the implication of these findings is that if the

APPENDIX TO CHAPTER 4

habitat in the south (Prairie Provinces and north-central United States) is depleted, migrating waterfowl will continue north in search of satisfactory nest sites rather than crowd onto the southern breeding grounds.[37] Although the water areas in the Dakotas, Minnesota, and Montana were not incorporated into this analysis because of noncomparable water counts, the trend in water areas in these regions complemented that in the Prairie Provinces, and consequently the results of the analysis are almost certainly applicable to the breeding grounds in the north-central United States as well.

[37] I wish to thank Walter F. Crissey for releasing his data to me so that I might estimate the above relations and perform this analysis.

5

THE DISTRIBUTION OF HUNTING LAND IN MINNESOTA

When the problem of the distribution of permanent wetlands in Minnesota was considered in chapter 2, it was assumed that these lands had no value as breeding habitat ($W = 0$), and that their rental price as a hunting preserve was zero ($R = 0$). One of the significant results of the analysis there was that if perfectly competitive farm prices prevailed, the investment cost of reclaiming permanent wetlands was prohibitively high. Hence, it was possible to infer, without any reference to values in the wildlife sector, that conversion of such marshes into arable land is an economically inefficient use of resources.

Optimality, then, is not at issue here. Of concern in this section is the allocation of resources under existing conditions. Since it is patently unrealistic to expect the immediate discontinuance of the agricultural price support program, it seems appropriate to consider the severity of the maldistribution of permanent marshes induced by crop subsidies.

It is conceivable that no such maldistribution of permanent wetlands occurs and that property owners will preserve their wetlands if hunters are willing to pay a high-enough price to obtain exclusive use of recreational hunting land. In chapter 2 it was also found that, given a zero rental value for hunting land, price supports for agriculture make drainage of permanent wetlands a profitable venture in southern Minnesota, but a marginal project at best in the northwestern part of the state. The question investigated here is whether the rents that marsh owners can currently obtain for hunting land are high enough to discourage reclamation. It is important to recognize, however, that the current rental value for hunting land cannot legitimately be accepted as a measure of the social value of such land. This price, though competitively determined, is arrived at in competition with access-free, public hunting preserves, and such competition undoubtedly acts to depress the rent that can be charged by private landowners.

Allocation of Privately Owned, Permanent Wetlands

In order to distinguish social from private values let

V_{wp} = the discounted value of the rental income from leasing an acre of hunting land, or the private value of an acre of marsh to the wetlands owner

$$= \sum_{t=1}^{n} \frac{R - C_d - C_w}{(1+r)^t},$$

where R is the price that an acre of hunting land currently commands in the market, and C_d and C_w are nuisance and maintenance costs, as defined previously. The assumption that C_d and C_w are zero is retained.[1] It will be seen later that this assumption strengthens the argument.

Let the discounted value of the income stream from an acre of drained, permanent wetland, when output is evaluated with subsidized prices be

$$V_{asp} = \sum_{t=1}^{n} \sum_{i=1}^{m} \frac{p_i q_i - C_i}{(1+r)^t},$$

where all the variables remain as defined previously. We are interested in the direction of the inequality in the following expression:

$$V_{wp} \lesseqgtr V_{asp} - I_s,$$

[1] See chapter 2, pp. 27 and 33 for the discussion of C_d and C_w. W does not appear in the expression for the private value of wetlands, since it is not a value that can be appropriated by the property owner.

i.e., we wish to compare the discounted value of the rental income from leasing an acre of hunting land with the difference between the present value of an acre of arable land and the cost of reclaiming that land. The right-hand side of the inequality is readily derivable from table 7 (p. 37) and our estimates of the cost of recovering permanent wetlands. $V_{asp} - I_s$ takes on alternative values depending upon the estimated capital outlay for drainage (high, low, or mean I_s) and the estimated revenues from support prices. The values appear in table 9.

TABLE 9. PRESENT VALUE OF AN ACRE OF ARABLE LAND MINUS THE COST OF RECLAIMING IT ($V_{asp} - I_s$)

Estimated net revenues at support prices	Southern Minnesota 20 years	Southern Minnesota 50 years	Northwestern Minnesota 20 years	Northwestern Minnesota 50 years
Low investment cost:				
Highest support revenues	$178.13	$254.30	$28.90	$71.91
Lowest support revenues	151.92	222.26	*	20.33
Mean investment cost:				
Highest support revenues	132.87	209.04	*	26.65
Lowest support revenues	106.66	177.00	*	*
High investment cost:				
Highest support revenues	38.31	114.48	*	*
Lowest support revenues	12.10	82.44	*	*

* $V_{asp} - I_s \leq 0$, making drainage unwarranted.

Define V'_{wp} as the present value for an acre of hunting land that just leaves the wetlands owner indifferent between draining and not draining. That is,

$$V'_{wp} = V_{asp} - I_s = R'\left[\frac{1 - (1 + r)^{-n}}{r}\right],$$

where R' is the value for R that just produces the equality of V'_{wp} and $V_{asp} - I_s$. Setting r at 8½ percent, and solving for R' gives the values shown in table 10—the rents necessary to forestall drainage under alternative assumptions about capital lives and revenues from agriculture.

TABLE 10. PER ACRE RENTAL VALUE OF WETLANDS NECESSARY TO FORESTALL DRAINAGE

| | Life of the asset ||||
| | Southern Minnesota || Northwestern Minnesota ||
Revenues	20 years	50 years	20 years	50 years
Low investment cost:				
Highest support revenues	$18.82	$21.99	$3.05	$6.22
Lowest support revenues	16.05	19.22	*	1.76
Mean investment cost:				
Highest support revenues	14.04	18.07	*	2.30
Lowest support revenues	11.27	15.30	*	*
High investment cost:				
Highest support revenues	4.05	9.90	*	*
Lowest support revenues	1.28	7.13	*	*

Note: The values are derived from data in table 7 and table 9 and from the estimates of the cost of recovering an acre of permanent wetlands (low, $164.75; mean, $209.98; and high, $304.54). Discount rate is 8½ percent.

* $V_{asp} - I_s \leq 0$. This means that drainage is unwarranted except for a special and unlikely situation were C_d is large enough to make V_{wp} less than $V_{asp} - I_s$.

Rental Value of Minnesota Hunting Land

In order to estimate the rental value of hunting land, a mail sample was conducted of Minnesota wetlands owners who were reported to have engaged in leasing the hunting rights to their land. The mailing list was compiled as follows: (1) all property owners who advertised their hunting sloughs for rent in the *Minneapolis Star* and *Minneapolis Tribune* during the period 1962–64 were included in the sample. (2) Every game warden in the state was requested, by mail, to supply the names of owners in his region who were potential lessors. (3) Waterfowl hunters who received the questionnaire discussed in chapter 3 were asked to supply the name and address of the lessor if they leased hunting land in 1964. These three sources produced 447 names, and a blanket mailing was made to all of them. The response statistics are shown on the next page.

Number of people in the sample	447
Nonrespondents	238
Questionnaire returned by Post Office	17
	255
Respondents	
Received rental income for access to land only	66
Received rental income but provided other services (dogs, blinds, lodging, boat, etc.) which could not be separated from the gross rental fee reported	22
Received no income	65
Do not own wetlands in Minnesota	22
Incomplete questionnaire	17
	192

With only 66 usable responses, any broad inference is a little hazardous. This would surely be the case if the distribution of the sample were dispersed. In this sample, however, it is hard to dismiss the large frequency of observations at low rental values. More than half the rental fees fall below $2.00 per acre, and over 70 percent lie below $4.00.[2] The complete distribution is shown below.

Rent	Frequency	
$ per acre		
0.00– 0.50	13	
0.51– 1.00	11	
1.01– 1.50	6	Median = $1.78
1.51– 2.00	6	
2.01– 2.50	4	Mean = $1.94
2.51– 3.00	4	
3.01– 4.00	4	
4.01– 6.00	5	
6.01– 9.00	8	
9.01–20.00	3	
20.01–40.00	2	
	66	

[2] I know of one other study of rental values of hunting areas in Minnesota: Hans G. Uhlig, "Survey of Leased Waterfowl Hunting Rights in Minnesota," *Journal of Wildlife Management*, vol. 25, no. 1 (1961), p.

A comparison of these data with the estimates of the rents necessary for retention of permanent habitat (table 10) suggests that few of the actual fees that hunters are willing to pay are sufficient to forestall drainage in the southern part of the state.[3] Some of the rental fees in the sample are large enough to discourage reclamation of those wetlands in the south which are most expensive to drain (table 10, high investment cost, 20-year life expectancy), but the median rental is far too low to prevent conversion of most of the permanent marshes in this region. The fact that very few of the rentals in the sample were in southern Minnesota tends to support this conclusion. It should be observed that the assumption that the nuisance cost associated with wetlands is zero strengthens the result here, for positive nuisance costs would only increase the incentive to drain.

As indicated in chapter 2, the expense of draining permanent wetlands in the northwest precludes reclamation in most instances, even with support prices in effect. For those potholes in northwestern Minnesota that can be cheaply converted to arable land (table 10, low investment cost), no universal conclusion can be inferred from the results of this sample. The average rent is $2.27 per acre for the western counties for which observations are available.[4] This rental is sufficient to prevent drainage of the less fertile lands in the northwest (table 10, lowest support revenues), but not of the relatively productive areas in this region (highest support revenues).

While crop price supports are in effect, almost all of the permanent wetlands in the south appear to be in jeopardy, but few of those in the northwest are threatened. However, I wish to emphasize again that the conclusions with respect to the perma-

---

204. Uhlig conducted his study in 1959, had a sample size of 49 (of which all were surveyed), and found a mean per acre rental fee of $5.10. It is not clear whether services other than access to the hunting area were provided by the lessors, for Uhlig loosely described the leases included in the sample as having the "primary purpose" of waterfowl hunting. It is of no consequence, however, for even this price is only sufficient to prevent installation in situations where drainage would be most costly.

[3] This should not be surprising in the light of the nonpriced competitive public hunting areas alluded to above.

[4] Polk, Clay, Becker, Wilkin, Otter Tail, Todd, Douglas, Grant, Stevens, Big Stone, Pope, Stearns, Kandiyohi.

nent wetlands of northwestern Minnesota are particularly suspect. Lack of confidence in these results emanates from the fact that it was assumed that both tile mains and a ditch would be required for adequate drainage in the northwest, whereas it is quite likely that for many projects a ditch will be sufficient. Where tile is not needed, the cost of reclamation will be lower, and the rent needed to prevent drainage will be higher than that estimated in table 10. The conclusions regarding permanent wetlands in the northwest would have to be tempered further should crop destruction by waterfowl or other nuisance costs prove to be more severe than the evidence suggests.

The Influence of Drainage on Hunting Quality

Having established that reclamation of some permanent wetlands seems likely, especially in southern Minnesota, it is important to determine the influence such drainage will have on the quality of hunting in the state. Since 15–25 percent of the birds killed in the state are taken in the south,[5] drainage of hunting areas here is potentially serious. However, it is quite difficult to assess with much precision the opportunity cost of a reduction in the stock of permanent wetlands in southern Minnesota. Drainage of wetlands in Minnesota has implications for hunting not just in the immediate area, but throughout the Mississippi Flyway as well.

The effect of depletion of domestic breeding grounds upon the waterfowl population in Minnesota was considered in chapter 4. Since many of the privately owned hunting sloughs serve also as breeding habitat, the results of that analysis are applicable here. We concluded in chapter 4 that if waterfowl breeding habitat in Minnesota (or in any of the other southern breeding grounds) were eliminated, migratory waterfowl would

[5] The counties used in calculating this estimate were Blue Earth, Cottonwood, Dodge, Faribault, Freeborn, Jackson, Le Sueur, Lyon, McLeod, Martin, Mower, Murray, Nicollet, Nobles, Redwood, Renville, Rice, Scott, Sibley, Steele, Waseca, Watonwan, Yellow Medicine. Southern Minnesota was defined as this set because these counties lie south of the Twin Cities, and because tile drainage was prominent in these areas. The estimate of the percentage of the state harvest shot here is based on the 1963 and 1964 Hunter Report Card Surveys conducted by the Minnesota Department of Conservation.

continue north in search of satisfactory nesting sites rather than crowd onto the reduced southern habitat. Thus, if the breeding grounds throughout the rest of Minnesota are at their carrying capacity, reclamation of permanent water in southern Minnesota is likely to have a significant impact upon the number of waterfowl nesting in Minnesota, and also upon the hunting harvest in Minnesota, for, as was seen in the preceding chapter, Minnesota resident birds have a much higher probability of being shot within the state than do birds whose nest site is elsewhere. Forcing birds to migrate to breeding grounds remote from Minnesota will reduce the hunting quality in the state.

If the breeding grounds in the Dakotas and southern Canada are also at their carrying capacity, then drainage of permanent water in Minnesota will have a depressing effect upon the size of the continental waterfowl population (and hunting quality) as well, because the displaced birds will be forced to move to northern Canada where they seem to reproduce less plentifully. Since only a small percentage of the continental population nests in southern Minnesota, the effect of drainage in this area is likely to be insignificant. Such drainage throughout all the southern breeding grounds would have a much more alarming impact, of course.

There are two additional reasons why continued depletion of permanent wetlands in southern Minnesota will diminish the quality of hunting in Minnesota. Fewer migratory birds will be attracted to Minnesota in the fall, and both hunters and hunted will be forced to crowd onto the remaining areas. Under crowded conditions, waterfowl tend to be a good deal safer.[6] In an effort to obtain their quarry before some rival hunter frightens the waterfowl, hunters tend to commit themselves too soon and shoot at birds that are out of range. The probability of killing a bird is increased if hunters permit the birds to land

[6] If the size of the migratory population were not significantly reduced by the drainage in Minnesota, then states further south on the Flyway might enjoy an improved hunting quality. Fewer birds would be bagged in Minnesota, and more birds would find their way to the southern states. However, since drainage is a widespread phenomenon in the Midwest, the size of the migratory population is likely to be affected, and hunting quality is likely to suffer throughout the Flyway.

on the marsh, and save most of their fire until the birds are rising after having been routed by the initial shots. Frequently, under conditions of crowding, a hunter is induced to fire prematurely as the waterfowl make their approach, because he is afraid that a competitor may shoot first, scattering the ducks before he can get off a shot. Thus, the problem has aspects not unlike that of crowding on highways, where a divergence between private and social costs occurs as the number of participants increases, and the time required to achieve one's goal is multiplied.

If an estimate of the additional time costs imposed upon hunters through crowding could have been made, then some estimate of the economic costs involved would have been possible, because an estimated income distribution for this group is available. But the problems involved in forecasting the time costs resulting from crowding, as well as the extent of crowding produced by drainage of permanent habitat, are formidable. Nevertheless it is clear that drainage of permanent water in southern Minnesota cannot but reduce the quality of hunting throughout the state.

Allocation of Publicly Owned Hunting Land

Mention was made at the beginning of this chapter that present rental values for privately owned hunting land could not be accepted as a measure of the social value of that hunting land, because they were arrived at in competition with entry-free, public hunting preserves. Free use of public hunting areas causes these areas to be used too intensively. It also depresses the rent for private land and thereby increases the likelihood of reclamation of such land.

The formal economic model in figure 2 demonstrates that public hunting land would be more efficiently used if an access fee were imposed upon hunters.[7] Assume that hunters seek

---

[7] The analysis here is heavily dependent upon H. Scott Gordon's generally applicable work, "The Economic Theory of a Common-Property Resource: The Fishery," *The Journal of Political Economy*, April 1954, pp. 124–42, and James A. Crutchfield, "Valuation of Fishing Resources," *Land Economics*, May 1962, pp. 145–48.

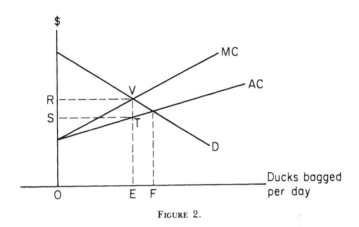

FIGURE 2.

ducks in the bag, and let D represent the aggregate demand for ducks per day at a particular hunting preserve. D is downward sloping reflecting the assumption that individual hunters have diminishing marginal rates of substitution between ducks and other commodities. The number of ducks bagged is a function of the amount of hunting effort (measured in number of hunter days) applied to the marsh. Let AC be the private opportunity cost to a hunter of bagging a duck. The analysis is enormously simplified by ignoring fixed costs (travelling time, guns, etc.) and by assuming a uniform private opportunity cost, and the results are not significantly altered. The assumption of a uniform opportunity cost is heroic but standard, and the mean income of hunters may be used to measure AC.

AC is upward sloping, because as additional hunters crowd onto the marsh, the time required for any one hunter to bag a duck increases. The presence of rivals induces individual hunters to undertake behavior in their own self-interest which reduces the probability of any one hunter bagging a duck during a given period of time. It is not necessary to assume that additional hunters destroy a significant portion of the waterfowl population resting in the region in order to achieve an upward sloping AC. An infinite bird population could be assumed, and AC would still possess a positive slope due to the behavioral assumption about hunters.

MC represents the social cost of bagging an additional duck. The divergence between AC and MC reflects the difference between private and social costs. The marginal hunter considers AC as the cost to him of acquiring an additional bird, ignoring the negative external influence which he exerts upon the productivity of his rivals. Assuming that each hunter is motivated by the desire to maximize the value of the ducks that he kills, hunters will patronize the preserve that yields the highest value (in terms of ducks bagged) for their hunting effort. This pattern of behavior results in individual areas being utilized up to the point where $AC = D$. Under such circumstances, no return accrues to the marsh itself.

The optimal amount of ducks which should be taken from a marsh is OE, not OF, for at OE the social value of the additional bird is identical with the social cost of obtaining it. In order to restrict usage to the optimal amount, a charge of RS must be imposed for each duck bagged,[8] resulting in a rent of $RSTV$ accruing to the hunting preserve. If it is impractical to charge on the basis of ducks bagged, an appropriate entry fee to the marsh will serve a similar purpose. If the present value of this income stream plus the value of the land as breeding habitat exceeds the purchase price of a preserve,[9] more capital should be invested in hunting land. The state should divest itself of such land holdings if the inequality runs the other way.[10]

Summary

1. The estimate of the investment cost involved in the drainage of permanent wetlands is sufficiently high to prevent reclamation if perfectly competitive farm prices prevail and marsh owners pay the full cost of drainage. This conclusion holds even if the rental value of wetlands for recreational purposes is zero. The nuisance value of the wetlands would have

---

[8] I am not advocating a uniform price for a bird at all preserves. Clearly, higher fees will be required in choice hunting areas.

[9] It is assumed that competitive prices for agricultural produce would be used in computing this purchase price.

[10] It should be noted here that no presumption in favor of public ownership is implied. An appropriate set of subsidies to private marsh owners could be established to effect the desired, optimal allocation of wetlands.

to be considerably higher than that indicated by any of the current evidence for this result to be reversed, and drainage to become economically efficient.[11]

2. If current farm price supports remain in effect, the actual fees paid by hunters to rent wetlands for recreational purposes are not sufficient to forestall drainage in the southern part of the state. In the northwest, however, the expense of draining permanent wetlands precludes reclamation in most instances, even with price supports in effect; even an inexpensive project would have to exhibit relatively high productivity to warrant drainage. Hence, as long as crop price supports are in effect, almost all of the permanent wetlands in the south are in jeopardy, while few of those in the northwest are threatened. One qualification is in order: if a ditch is sufficient to drain permanent wetlands in the northwest, or if nuisance costs are somewhat higher than estimated, then many more of these water areas are threatened with reclamation than the analysis indicates.[12]

3. Since the amount of hunting in southern Minnesota is considerable (between 15 percent and 25 percent of the duck harvest being reaped in this region), the effect of reclamation in the south on hunting quality cannot be easily dismissed. Estimation of the economic loss involved, however, is an unsolved problem. Continued depletion of permanent wetlands will result in both hunters and their quarry crowding onto the remaining areas in the fall, a smaller number of migratory birds will probably be attracted to the area in both the spring and fall, and fewer birds will be bred and brought to flight stage in Minnesota. All of these effects tend to reduce hunting quality in Minnesota.

4. In the absence of a user charge (entrance fee) for public hunting preserves, public hunting land is used too intensely.

[11] See chapter 2, footnote 27, p. 36.
[12] Nuisance costs would have to be almost $15 an acre annually in order to make drainage socially appropriate in this area for a project of mean expense (mean I_s). But they would only have to be between $2 and $6 to make such a project marginally profitable under conditions of price supports for agriculture.

Policy Implications

If wetland resource misallocations induced by subsidies are judged to be socially undesirable, then some program to prevent the continued erosion of the stock of both permanent and temporary wetlands is necessary. The reclamation of permanent wetlands and many temporary wetlands would not continue if subsidies were not provided to the agricultural sector and if competitive prices existed in the market for agricultural commodities.[13] With price supports in effect, however, drainage becomes privately profitable despite the fact that it is socially inefficient.

There are several possible ways of dealing with the problem. Terminating price supports would correct much of the problem, but this is not a feasible alternative politically. The price support program is well entrenched and has withstood attacks based on more compelling criticism than that offered by the results of this study. However, it might be possible to retain the income redistributive features of the support program without providing incentives to engage in wetland resource misallocations. This could be done by eliminating payments that encourage socially inefficient drainage and substituting payments to farmers for maintaining wetlands that are judged to be important for wildlife habitat. In fact, such payments (called easement payments) are currently authorized under state and federal programs, but these programs are underfunded and unwieldly and do not provide adequate means to significantly retard the rate of agricultural reclamation of wetlands.

A more vigorous program along these lines might be undertaken and could be financed from a number of revenue sources. One source (and probably the most feasible one politically) would be from a tax levied on duck hunters. It might seem inequitable to single out duck hunters and ignore others, such as trappers, photographers, naturalists, and so on, who also obtain direct benefits from wetlands, but administrative difficulties are likely to rule out a more broadly based beneficiary tax. However, it might be possible to finance an easement payments

---

[13] See chapter 2, pp. 39–40, and chapter 4, pp. 84–85.

program out of general government revenues if the payments were viewed as income maintenance payments (in the same sense that price supports are) as well as a means for promoting allocative efficiency. The program could be administered by the Department of Agriculture just as the agricultural support program is.

Another way of dealing with the problem would be to prohibit drainage without a permit from the Fish and Wildlife Service. Requests for permission to reclaim land would have to be made to the FWS, which would evaluate the ecological characteristics of the land to determine its importance as wildlife habitat. Permits would be granted only for land with little or no value as wildlife habitat. A mechanism for evaluation is already provided under Public Law 87-732, requests for financial assistance in defraying drainage costs being referred initially to the FWS for approval.[14] The same administrative machinery could be enlisted to consider applications for permits to install drainage facilities. Building permits, pollution control measures, enforced compliance with zoning restrictions, and wilderness preservation legislation stand as ample precedents for such restraints where adverse third party effects are encountered. Similar means might be considered in connection with drainage activity.

Clearly, altering present policies so as to get an efficient allocation of wetlands will require sacrifices by some parties, and there is no scientifically established way to determine who is to bear the burden of the change. The income redistributive consequences associated with an economically determined change are an ethical matter. For example, taxation of hunters to transfer funds to marsh owners effects a redistribution of income that would not have occurred under perfectly competitive conditions in the agricultural sector. Much of the drainage would not have been feasible under perfectly competitive conditions, and hence no payment from hunters to farmers would have been necessary to forestall the reclamation of wetlands. Nonetheless, these are not sufficient grounds on which to reject a pro-

[14] See p. 12 for a description.

posal for easement payments. Just because such a transfer would not occur under perfectly competitive conditions does not mean that ethically it is inappropriate. Many if not all thoughtful members of society would reject the proposition that the distribution of income between farmers and the rest of society, including waterfowl hunters, would be socially sanctioned under competitive conditions in agricultural commodity markets.[15] Indeed, if there are some uncompensated costs to farmers for the maintenance of migratory waterfowl facilities, such as depredation of crops, then perfectly competitive markets would not reflect true social costs and gains, and there would be an implicit transfer from farmers to hunters.

On the other hand, a legal prohibition of drainage would subject farmers to possible capital losses because the value of a farm is likely to decline if wetlands can no longer be converted to cropland. A restriction on drainage would impose a real hardship on those who invested in large complexes of wetlands with the view of reclaiming them, but it is doubtful that such investors make up a significant proportion of the landowners whose wetlands are important as migratory waterfowl habitat. Farmers who have wetlands on their property or cropland located near waterfowl habitat may be subject to other costs, in addition to capital losses. A prohibition on drainage would force farmers to finance the breeding and some of the feeding of waterfowl and the maintenance of rest areas for migratory birds without providing any means by which they could reap the social benefits of their resource. They are unlikely to have any way of selling the product of their land, and they would be prevented from eliminating the source of the external diseconomies. In any event, if inequities resulted, wetland owners could be compensated for their losses by specific taxes levied

[15] One can say this, of course, without endorsing agricultural price supports as a desirable or even an effective way of accomplishing the income maintenance objective in the agricultural sector. For an analysis of the perverse nature of the income distribution associated with price supports see James T. Bonnen, "The Distribution of Benefits from Cotton Price Supports," in *Problems in Public Expenditure Analysis*, ed. Samuel B. Chase, Jr. (Washington: The Brookings Institution, 1968).

against beneficiaries of the wetlands preservation program or, if that is infeasible, from general revenues.

Although some reclamation of wetlands may be economically justified even from a social viewpoint, there appears to be no economic justification for granting public assistance for its accomplishment. Subsidies could be justified only if there were external economies associated with such drainage undertakings. No significant external economies, however, have been identified with drainage activity. On the contrary, if reclamation is conducted on a large enough scale, it may impose substantial flood costs on communities downstream from the drained lands. Occasionally drainage subsidies are defended as income redistributive devices that improve the circumstances of low-income families. This argument is without merit, since income is not a criterion considered by the authorities in determining eligibility for subsidy.